Lecture Notes in Electrical Engineering

Volume 552

The book series *Lecture Notes in Electrical Engineering* (LNEE) publishes the latest developments in Electrical Engineering - quickly, informally and in high quality. While original research reported in proceedings and monographs has traditionally formed the core of LNEE, we also encourage authors to submit books devoted to supporting student education and professional training in the various fields and applications areas of electrical engineering. The series cover classical and emerging topics concerning:

- Communication Engineering, Information Theory and Networks
- Electronics Engineering and Microelectronics
- Signal, Image and Speech Processing
- Wireless and Mobile Communication
- Circuits and Systems
- Energy Systems, Power Electronics and Electrical Machines
- Electro-optical Engineering
- Instrumentation Engineering
- Avionics Engineering
- Control Systems
- Internet-of-Things and Cybersecurity
- Biomedical Devices, MEMS and NEMS

For general information about this book series, comments or suggestions, please contact leontina. dicecco@springer.com.

To submit a proposal or request further information, please contact the Publishing Editor in your country:

China
Jasmine Dou, Associate Editor (jasmine.dou@springer.com)

India
Swati Meherishi, Executive Editor (swati.meherishi@springer.com)
Aninda Bose, Senior Editor (aninda.bose@springer.com)

Japan
Takeyuki Yonezawa, Editorial Director (takeyuki.yonezawa@springer.com)

South Korea
Smith (Ahram) Chae, Editor (smith.chae@springer.com)

Southeast Asia
Ramesh Nath Premnath, Editor (ramesh.premnath@springer.com)

USA, Canada:
Michael Luby, Senior Editor (michael.luby@springer.com)

All other Countries:
Leontina Di Cecco, Senior Editor (leontina.dicecco@springer.com)
Christoph Baumann, Executive Editor (christoph.baumann@springer.com)

** Indexing: The books of this series are submitted to ISI Proceedings, EI-Compendex, SCOPUS, MetaPress, Web of Science and Springerlink **

More information about this series at http://www.springer.com/series/7818

Liheng Wang · Yirong Wu ·
Jianya Gong
Editors

Proceedings of the 5th China High Resolution Earth Observation Conference (CHREOC 2018)

Springer

Editors
Liheng Wang
China Aerospace Science and Technology
Corporation
Beijing, China

Yirong Wu
Aerospace Information Research Institute
Chinese Academy of Sciences
Beijing, China

Jianya Gong
Wuhan University
Wuhan, Hubei, China

ISSN 1876-1100   ISSN 1876-1119   (electronic)
Lecture Notes in Electrical Engineering
ISBN 978-981-13-6552-2   ISBN 978-981-13-6553-9   (eBook)
https://doi.org/10.1007/978-981-13-6553-9

Library of Congress Control Number: 2019931537

This Springer imprint is published by the registered company Springer Nature Singapore Pte Ltd.
The registered company address is: 152 Beach Road, #21-01/04 Gateway East, Singapore 189721, Singapore

Contents

Satellite Energy Prediction and Autonomous Management Technology

Dongsheng Jiang[✉], Peng Tian, and Pei Zhang

Beijing Institute of Spacecraft System Engineering, Beijing 100094, China
Jiang_dongsheng@sohu.com

Abstract. The new generation satellite, especially the agile earth observing satellites have autonomous mission scheduling ability, which make the mission scheduling more complex and ask power system can be automate managed. This paper introduces a new autonomous management system of satellite energy. The system consists of two parts: on-board power self-management system and ground health management system. It is a satellite-ground integrated energy management system structure. This paper focuses on the two key technologies of the satellite energy autonomous management: energy prediction and estimation technology and energy autonomous management technology.

Keywords: Satellite · Energy prediction · Autonomous management

1 Introduction

Electrical power system is one of the most important subsystems of spacecraft, of which the main functions are to supply power to the whole satellite, using solar arrays absorb solar radiation and generate power during the light period. Besides that, it also charges and maintains the batteries. In the shadow period, the power of the battery is adjusted and controlled to provide a stable power supply voltage to the entire satellite [1]. Whether the satellite subsystem can ensure reliable and abundant supply of energy directly determines the normal operation of the equipment and the successful completion of the flight mission during the operation of the spacecraft.

With the various of satellite payloads and the complication of flight missions, especially the new generation of the satellites represented by agile earth observation satellites, due to the complex observation model, the satellite system proposes the power subsystem to provide energy prediction ability in the future, and support the mission planning and scheduling to satisfy the needs of satellite multitasking. At the same time, with the increasing number of satellites in orbit, the workload and work pressures of the people that work on ground-based satellites monitoring and control are both increasing. The satellite power subsystems are required to have certain capabilities of autonomous management and energy mode control. Current satellites and spacecraft, especially deep space exploration spacecraft, have long delays in communication. For example, Mars Rover's telemeter and control information are delayed about more than 40 min, therefore, independent energy prediction, energy scheduling, and fault management, fault isolation, and fault recovery capabilities are required.

© Springer Nature Singapore Pte Ltd. 2019
L. Wang et al. (eds.), *Proceedings of the 5th China High Resolution Earth Observation Conference (CHREOC 2018)*, Lecture Notes in Electrical Engineering 552,
https://doi.org/10.1007/978-981-13-6553-9_1

2   The Design of Satellite Energy Autonomous Management System

Beijing Institute of Spacecraft System Engineering has taken the lead in researching on spacecraft energy autonomous management technologies, and conducted in-depth studies on energy prediction technologies, energy dynamic scheduling programs, and power subsystem fault detection and isolation. The project is based on a new generation of satellites, which can meet the need for on-orbit power subsystems management of future high, medium, low orbit satellites and deep space exploration and moon

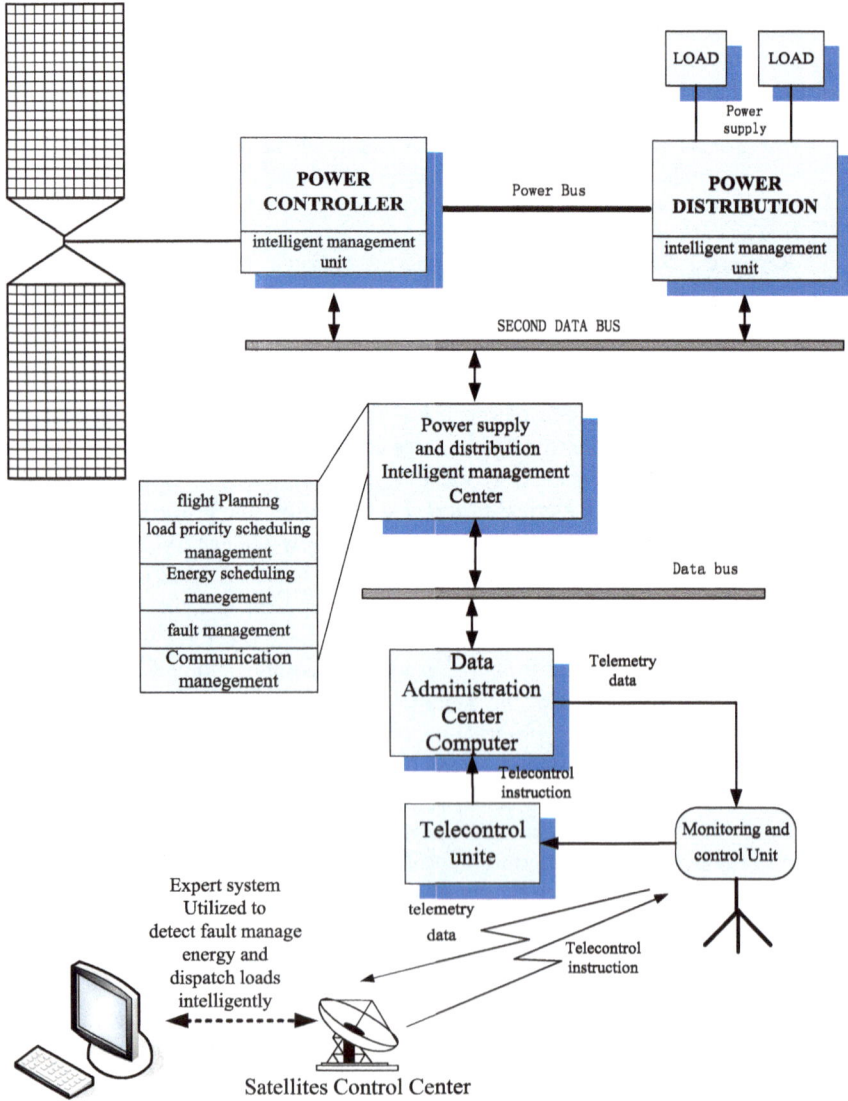

Fig. 1. Topology diagram of autonomous management power system

basement in energy prediction, energy dynamic scheduling, fault monitoring, fault isolation, and system reconstruction management as well as on-orbit healthy operation.

The spacecraft energy autonomous management system adopts a two-stage distributed computer system. It consists of an intelligent management unit for power distribution and multiple subordinate machines embedded in devices such as power distribution units and power control unit, and transmits telemetry data, status information, and instructions through a serial data buses shown in Fig. 1. The satellite energy autonomous management system includes two key technologies: energy prediction estimation technology and energy autonomous management technology.

2.1 Solar Array Output Power Prediction

The satellites solar arrays output power prediction model is a prerequisite for autonomous management activities such as power system fault warning, energy planning, and scheduling. It is mainly affected by parameters such as solar incident angle, solar distance factor, solar arrays temperature, and solar arrays attenuation factor. The solar incident angle θ and the solar distance factor Fd could be deduced from the satellite orbital numbers, and the solar incident angle variation trend and the solar distance factor are obtained as well.

The temperature T of the solar array is estimated by using the thermal balance equation of the solar array in combination with the solar radiation energy incident on the solar array, the Earth's solar reflective energy, the Earth's radiant energy, and the solar array's own thermal performance parameters.

Based on the actual equivalent circuit model of the solar cell, the I-V curve equation of the solar cell is derived as follows:

$$I = I_{SC}(1 - C_1\{\exp[V/(C_2 V_{OC})] - 1\}) \tag{1}$$

In the formula:

$$C_1 = [1 - I_{mp}/I_{SC}]\exp[-V_{mp}/(C_2 V_{OC})] \tag{2}$$

$$C_2 = [V_{mp}/V_{OC} - 1][\ln(1 - I_{mp}/I_{SC})]^{-1} \tag{3}$$

It can be seen from the above equation that only the four characteristic parameters of the solar cell, short-circuit current I_{sc}, the open-circuit voltage V_{oc}, the maximum power point current Imp, and the maximum power point voltage V_{mp} need to be input to determine the solar cell output current and voltage function (I-V curve). Solar array can also use this equation.

Solar cells are connected in series and in parallel to form a solar array. Under the condition of neglecting the loss of solar cell interconnections and the difference of solar cells, the output short-circuit current of the solar array composed of N_S series and N_P parallel is $N_P I_{SC}$, and the output open-circuit voltage is $N_S V_{OC}$, the maximum power point voltage is $N_s V_{mp}$, and the maximum power point current is $N_p I_{mp}$.

Through the ground radiation test, the degradation factor of the output voltage and the current of the solar cell on orbit can be obtained. Under the influence of the radiation, the solar cell open-circuit voltage V_{OC}, the short-circuit current I_{SC}, the maximum power point current Imp, the voltage V_{mp} change over time, and the I-V curve equation of the solar array under the influence of space radiation at any time in the orbit can be obtained.

For the on-orbit value of the solar arrays I_{SC}, V_{OC}, I_{mp}, V_{mp}, if the parameter of the solar cell slice AM_0 standard used is I_{SC0}, V_{OC0}, I_{mp0}, V_{mp0}, the I_{SC}, V_{OC}, I_{mp}, V_{mp} value of the on-track time parameter value is:

$$V_{OC} = [V_{OC0} + \beta_v(T-25)]K_{VRad}K_{VA}K_{Vuv} \cdot N_S \tag{4}$$

$$V_{mp} = [V_{mp0} + \beta_v(T-25)]K_{VRad}K_{VA}K_{Vuv} \cdot N_S \tag{5}$$

$$I_{SC} = [I_{SC0} + \beta_i(T - 25)]K_{IRad}K_{IA}K_{Iuv} \cdot N_P \tag{6}$$

$$I_{mp} = [I_{mp0} + \beta_i(T-25)]K_{IRad}K_{IA}K_{Iuv} \cdot N_P \tag{7}$$

In the formula:

β_v, β_i—voltage/current temperature coefficient, V/°C, A/°C;
T—operating temperature, °C;
K_{VRad}, K_{IRad}—voltage/current irradiation attenuation factor;
K_{VA}, K_{IA}—voltage/current connected loss factor;
K_{Vuv}, K_{Iuv}—voltage/current ultraviolet loss factor;

Substitute the value of I_{SC}, V_{OC}, I_{mp}, V_{mp} into the formula (1), (2), (3), we would get the I-V curve equation $I = f(V)$, and the actual operating point voltage is:

$$V_{work} = V_{bus} + V_{diode} + V_{wire} \tag{8}$$

In the formula:

V_{work}—solar arrays actual operating point voltage;
V_{bus}—bus voltage;
V_{diod}—isolation diode pressure drop;
V_{wire}—wire pressure drop;

Substitute the actual operating point voltage V_{work} into the I-V curve equation, we could obtain the operating point current I_{work}, and the actual output power P of the solar array can also be obtained [2]:

$$P_{work} = V_{works} \cdot I_{works} \cdot F_d \cdot \cos\theta \tag{9}$$

2.2 Energy Autonomous Management

For a spacecraft energy system, the amount of energy it can provide is limited. The main purpose of the early spacecraft power budget design was to ensure that the energy provided by the energy system was sufficient to meet the maximum energy demand envelope of the spacecraft's on-orbit mission planning and to ensure the success of the mission.

With the development of agile earth observation satellites and their demand for complex observation missions, the power subsystem cannot be designed to scale enough to satisfy the envelope of the maximum energy demand, so only mission energy dynamic planning, time-sharing, and more can be used. The multi-circle energy balance ensures the successful completion of the observation mission. The energy dynamic planning and scheduling process are shown in Fig. 2.

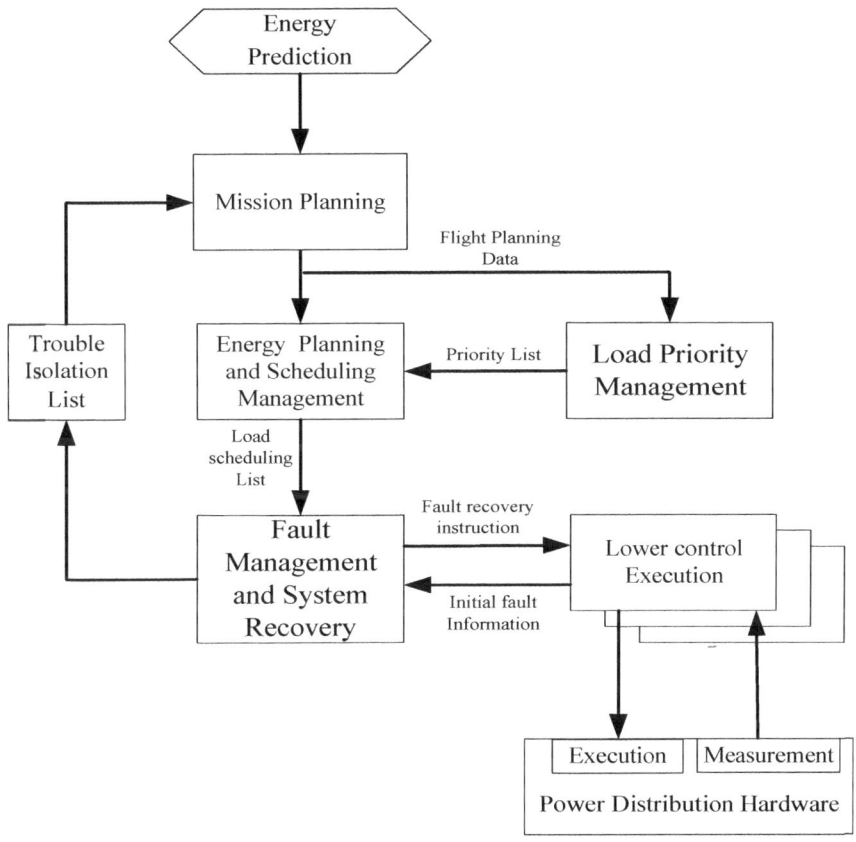

Fig. 2. Satellite power autonomous management procedure

The dynamic plan dispatch management of energy can be divided into two categories: macro-scheduling and micro-scheduling. The micro-scheduling strategy is to use one orbital cycle or one day of the satellite as a scheduling cycle. Its purpose is to propose a set of power supply protection rules that can immediately remove the lowest priority load, reduce the power consumption, and prevent the battery from over-discharging or over consumed due to non-isolated, the short-circuit fault in order to manage the load power supply and distribution of ensuring platform safety during the flight mission. Micro-scheduling usually adopts a rule-based scheduling strategy to manage load and power distribution. The load priority scheduling management method is usually used to provide power supply capabilities. First, the load with the least relation to the mission is removed step by step to ensure the base requirement of mission completion. When it is still overloaded, it will enter the minimum energy mode, which only maintains the basic operating load of the satellite platform and keeps the minimum use of electricity to assure for fault diagnosis and recovery [3].

The macro-scheduling strategy is based on one year or one flight mission cycle of a spacecraft as the scheduling cycle. Its purpose is to propose a set of energy optimization and utilization algorithms that can meet the contradictions between the planned electricity demand of the mission and the actual power supply capacity. The most significant goal is that make mission be completed. Besides that, to increase the lifetime of the orbit (increasing battery life) and maximize the optimal use of electrical energy is another goal. Therefore, long-term scheduling could be realized by macro-scheduling strategy.

Energy scheduling is a linear planning problem. The usual solution to this type of problem is to establish a corresponding optimal scheduling model and find the optimal solution. For spacecraft energy scheduling, energy use is usually maximized to ensure the completion of missions in each phase. The specific description is as follows:

Objective function: $f(S) = \min \sum_{t=1}^{T} Rs_t$

among them, $Rs_t = F_t/S_t$, $t = 1, 2, \ldots, T$

Energy balance constraint: $X_t = X_{t-1} + S_t - L_t - F_t$

Battery Constraints: $X_{\min} \leq X_t \leq X_{\max}$

Solar array power supply constraints: $\begin{cases} S_{t\min} \leq S_t \leq S_{t\max} \ldots \text{the light} \\ S_t = 0 \ldots \text{the shadow} \end{cases}$

Power consumption constraints: $\begin{cases} L_{\min} \leq L_t \leq S_t \ldots \text{the light} \\ L_{\min} \leq L_t \leq |X_t - X_{t-1}| \ldots \text{the shadow} \end{cases}$

In the formula, L_t is the power consumption of the load in t period, S_t is the solar array power supply capacity in t period, $S_{t\min}$, $S_{t\max}$ is the minimum and maximum power supply capacity of the solar wing in t period, X_{t-1} and X_t are the battery at the beginning of t period energy storage at the end of the period, $X_{\min}$ and $X_{\max}$ are the minimum and maximum storage energy (maximum discharge capacity and capacity) of the storage battery, F_t is the electrical energy diverted during period t, and Rs_t is the electrical energy loss rate during t period.

Complex mission spacecraft energy distribution and scheduling involve complex and combinatorial optimization solutions, which are non-monotonic and discrete. When using the dynamic planning method to solve, with the increase in the size of the mission, to find the optimal solution in the continuous space, there will often be dimension disasters. At present, the gradual optimization algorithm, genetic algorithm, and ant colony algorithm are usually used to solve the optimization scheduling problem, therefore we could quickly and accurately seek the best solution.

2.3   Space-Ground Integration Energy Autonomous Management System

Build a set of space-ground integration energy autonomous management system, including on-board power autonomous management system and ground health management system.

(1) on-board power autonomous management system

On-board power autonomous management system, including two control loops: one is a programmable control solid switch modules of power distribution, equipped with the protection of over current and over voltage circuit used to detect the troubles on the over current and over voltage fault of power-supply, achieve the quick judge and trouble isolation. The other one is an on-board computer control and management loop, which includes the intelligent troubles detection, system control and load distribution control by set priorities. This system integrates an expert system based on troubles model and operation rule, and can be utilized to detect fault manage energy and distribute loads intelligently. If the troubles might threaten the satellite's safety, this system would carry through autonomous diagnosis and emergency processing. Besides that, combining with the real flight mission, it would achieve on-board loads switch off and short-term energy scheduling.

(2) ground health management system

Due to the restriction of the computing ability and store space of on-board computers, and because of the limit of flexibility and adaptation of on-line trouble handling algorithm, therefore, power autonomous management system must be equipped with ground assistant trouble-decision management system—a set of expert system. This system utilizes the on-board telemetry data, such as trouble parameters, to analysis and estimate on-board troubles, and monitor the health status of on-board electrical power system. It also can assistance for the trouble prediction, the trouble diagnosis and the trouble handling, in the final, achieve the long-term management of on-board power. Compare the ground system with the on-orbit system, the ground system is not limited by the space, calculation resource and time, besides that, we also could equip the simulation system, confirmation system, database system and so on, and basing on above conditions, the ground expert system would give us an accurate diagnosis of on-board troubles.

For long-term on-orbit operational satellites, the slow degraded parameters are one of the main reasons that would influence and make a decision the age of satellites. Fortunately, the most of slow degraded parameters is predictable, and the aim of us modeling is that we use the expert information and occurred trouble case to gain the characteristic of trouble, gain the trouble model by statistics method, mathematics method, information fusion, artificial intelligence and so on, and in the end, we could build knowledge base of trouble model.

3 Conclusion

Energy autonomous management technology of satellites could improve it's autonomous operation ability, and let the computer of satellites accomplish energy prediction, energy supply and load distribution by itself to achieve the dynamic balance between the energy supply and the load requirement. It can help satellites using the energy more sufficiently and make flight mission scheduling and autonomous scheduling of satellites come true. In this paper, we introduce the output power of solar array prediction and estimation technology and energy autonomous management technology in detail, besides that, we provide a method of power autonomous management system on board as well.

References

1. Ma Shijun. Satellite power technology [M]. Beijing: China Astronautics Press, 2001 (in Chinese).
2. Jiang Dongsheng, Chen Qi, Zhang Pei, Study on Intelligent Management Technology for Spacecraft Electrical Power[J];Spacecraft Engineering, 2012, 21(4):100–105 (in Chinese).
3. Jiang Dongsheng Liu Shuhao Ma Ning. The Tools and Method for Satellite Solar Array Power Prediction on Orbit [C]// The 4th China High Resolution Earth Observation Conference, 2017, 8.

Research on Key Technologies
of Geosynchronous SAR Imaging System

Xiaoming Zhou[1]([⊠]), Dejin Tang[2], Caiping Li[1], and Linzhe Lao[1]

[1] Beijing Institute of Remote Sensing Information, Beijing 100192, China
zxm2913@163.com
[2] National Geomatics Center of China, Beijing 100830, China

Abstract. Aiming at the high altitude and long synthetic aperture time of the geostationary orbit SAR (GEOSAR) system, this paper has the characteristics of the different imaging model algorithm with the low orbit spaceborne SAR (LEOSAR), and analyzes its unique system characteristics in detail. The GEOSAR space geometric model, echo signal model, imaging algorithm, and process are studied and introduced. An optimal time domain compensation scheme is used to change the nonlinear time–frequency relation of the signal and weaken azimuth denaturation, and then combined with the frequency domain algorithm of two-dimensional NCS, the large scene image processing is carried out, and the experiment simulation is carried out. The research content has certain applicability and can be used as a basic theoretical research achievement for subsequent test stars.

Keywords: Microwave remote sensing · GEOSAR · Time–frequency compensation · Large scene imaging

1 Introduction

The Geosynchronous Earth Orbit SAR (GEOSAR) runs at the height of 36,000 km, with an hour level revisit capability and a large scene coverage. Compared to the ordinary low orbit (LEO) satellite, the geostationary orbit satellites are running for 24 h a week because of high orbital height. It is far larger than the LEO satellite [1]. In addition, the trajectories of the satellites in the earth's synchronous orbit are reclosing, and the LEO satellites usually have different trajectories per circle, so the advantage of the earth's synchronous orbit is that the satellite of the earth's synchronous orbit has a shorter revisit time than the LEO satellite. Compared with LEO satellites, GEO SAR needs longer time to achieve the same synthetic aperture angle, that is, the synthetic aperture time of GEO SAR is much larger than that of LEO satellites. Long synthetic aperture time leads to the conventional equivalent strabismus velocity model that cannot accurately characterize the motion trajectory of GEO SAR, so the traditional LEO SAR imaging algorithm is not suitable [2, 3]. Therefore, the GEOSAR imaging theory involves extensive and complex content research, including the GEOSAR spatial geometry model, the trajectory characteristic of the star point, the speed and range of beam scanning, the influence factors of the Doppler frequency and the synthetic aperture time, the establishment of the signal distance model, the design of

© Springer Nature Singapore Pte Ltd. 2019
L. Wang et al. (eds.), *Proceedings of the 5th China High Resolution Earth Observation Conference (CHREOC 2018)*, Lecture Notes in Electrical Engineering 552,
https://doi.org/10.1007/978-981-13-6553-9_2

imaging algorithm and the subsequent image interpretation, etc. This article briefly introduces and summarizes the above contents [4, 5].

2 GEOSAR Geometric Model

2.1 GEOSAR Echo Signal Model

The geometric model of GEO SAR is shown in Fig. 1. Because the synthetic aperture time of GEO SAR is far greater than that of LEO and airborne SAR, and the speed and acceleration of the satellite in GEO SAR vary greatly, the traditional equivalent velocity skew model cannot accurately characterize the oblique distance of GEO SAR, and a new oblique distance model must be adopted to approximate. The position vector of the satellite is assumed to be $\vec{r}_{sn}(x_s(t_a), y_s(t_a), z_s(t_a))$, the t_a is azimuth time, the position vector of the target A is $\vec{r}_{gn}(x, y, z)$. The distance between the radar and the target can be expressed as:

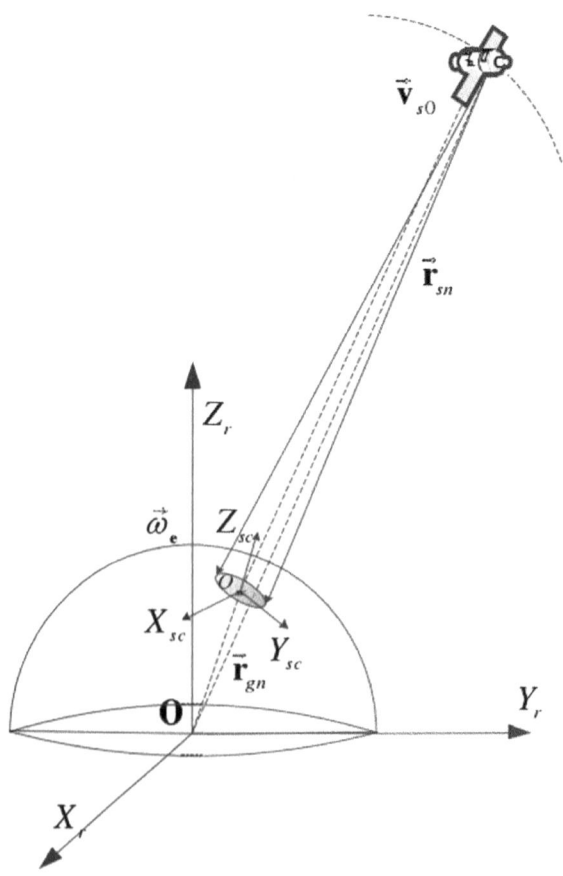

Fig. 1. GEOSAR geometric model

$$R(t_a) = \left\| \vec{r}_{sn} - \vec{r}_{gn} \right\| \tag{1}$$

The above formula can be obtained by Taylor expansion.

$$R(t_a) \approx R_0 + k_1 t_a + k_2 t_a^2 + k_3 t_a^3 + k_4 t_a^4 + \cdots \tag{2}$$

In theory, the formula can be expanded to an arbitrary order by Taylor, but in fact, in the light of the calculation complexity and the actual needs in the imaging process, the oblique distance model is generally only expanded to the finite order.

$$R(t_a) \approx R_0 + k_1 t_a + k_2 t_a^2 + k_3 t_a^3 + k_4 t_a^4 \tag{3}$$

2.2 GEOSAR Echo Two-Dimensional Frequency Domain Expression

Based on the above slant distance model, for the point target t_a with the shortest slanting distance of R and direction time, the expression of the echo signal is:

$$F(t_r, t_a; R) = \exp\left[j\pi k_r \left(t_r - \frac{2 \cdot R(t_a; R)^2}{c} \right) \right] \exp\left[-j\frac{4\pi R(t_a; R)}{\lambda} \right] \tag{4}$$

The distance to Fourier transform is used to deduce the expression of the azimuth time–frequency signal in the range frequency domain signal.

$$F(p_r, t_a) = -\frac{j\pi p_r^2}{k_r} - 4\frac{j\pi (R_0 + k_1 t_a + k_2 t_a^2 + k_3 t_a^3 + k_4 t_a^4)(p_r + p_c)}{c} \tag{5}$$

Due to the use of the four order skew model, the conventional method of deriving the stationary phase point needs to solve the three equation, and the solution is too complex. Therefore, this paper derives the stationary phase point from the way of series reversal. Finally, the two-dimensional frequency domain expression of the echo signal is obtained.

$$
\begin{aligned}
F(p_r, p_a) = \\
& -\frac{\pi f r^2}{kr} - \frac{4\pi R_0 (p_r + p)}{c} \\
& + \frac{\pi c}{4k_2(p_r + p_c)} \left(p_a + \frac{2k_1}{c} \cdot (p_r + p_c) \right)^2 \\
& + \frac{\pi k_3 c^2}{16k_2^3(p_r + p_c)} \left(p_a + \frac{2k_1}{c}(p_r + p_c) \right)^3 \\
& + \frac{\pi (k_3 c^2)}{16k_2^3(p_r + p_c)^2} \left(p_a + \frac{2k_1}{c}(p_r + p_c) \right)^4
\end{aligned}
\tag{6}
$$

3   GEOSAR Image Processing Algorithm for Large Strabismus

In order to improve the revisit performance of earth observation, GEO SAR often needs to work in a large squint mode. This part mainly studies the GEO SAR large strabismus imaging processing, and uses the optimal time domain compensation scheme to change the nonlinear time–frequency relation of the signal and weaken the azimuth denaturation, and then combined with the frequency domain algorithm of two-dimensional NCS for large scene imaging processing.

The processing algorithm of GEO SAR large squint imaging algorithm is shown in Fig. 2, which consists of two parts: optimal time domain compensation and two-dimensional NCS. The optimal time domain compensation is to achieve range walk correction and weaken the two and three term null variation characteristics of the original signal. The optimal time domain compensation is used to compensate the original echo in time domain.

$$H_{tc} = \exp\left[j\frac{4\pi}{\lambda}\left(k_{10}t_a + a_{op}t_a^2 + bt_a^3\right)\right] \tag{7}$$

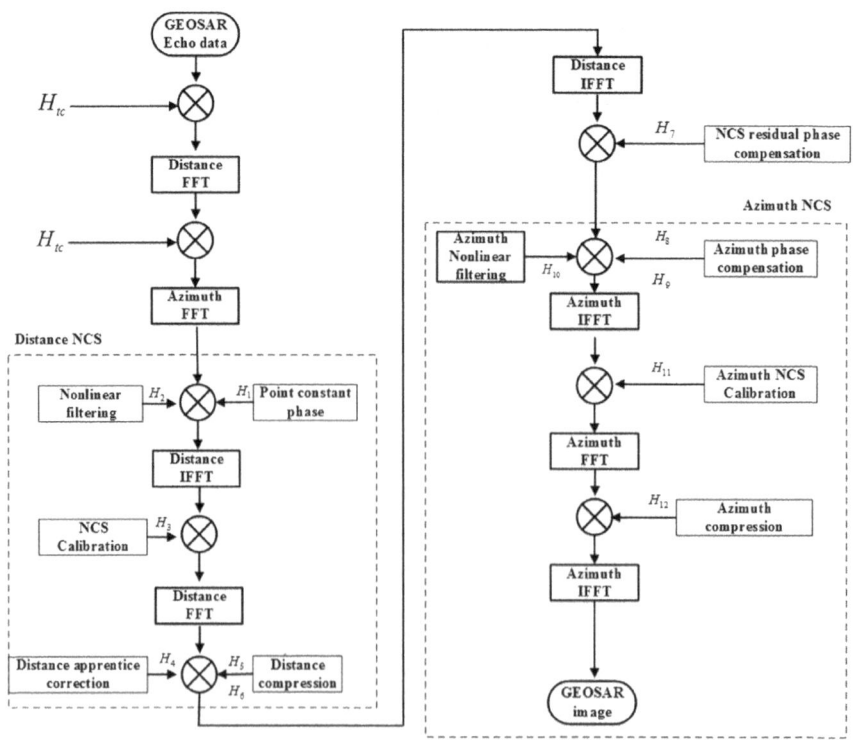

Fig. 2. Imaging algorithm flow

Among them, the first coefficient of the central point of the original scene k_{10} is carried out to compensate for the walk, and the last two items of compensation calculated a_{op} by b are the three times proportional component of the compensation. At the same time, in order to achieve consistent envelope and phase, envelope compensation is also needed.

$$H_{tc} = \exp\left[j\frac{4\pi}{c}\left(k_{10}t_a + a_{op}t_a^2 + bt_a^3\right)p_r\right] \tag{8}$$

After the optimal time domain compensation for the original echo, the azimuth denaturation of the original signal is weakened, and the range walk correction is realized. Then, a new range and azimuth denaturation calculation are carried out for the compensated signal. Then, the two-dimensional NCS frequency domain algorithm is used for focusing processing to realize the large scene imaging.

4 Simulation Experiment

In order to verify the effectiveness of the algorithm, a 300 km × 300 km lattice target simulation is carried out in this paper. At 0° latitudes, points are arranged at intervals of 75 km, and 25 points are arranged. The simulation parameters are shown in Table 1, and the imaging results are shown in Fig. 3. In order to verify the effectiveness of the algorithm, the four-point A–D on the edge angle and the scene center point E are plotted, then the PSLR and ISLR are evaluated, and the results are shown in Table 2. From the point of view of the table, the focusing effect of the point target is good, and the effectiveness of the imaging algorithm proposed in this chapter is verified.

Table 1. GEOSAR imaging simulation parameters

Parameter	PRF	Bandwidth	Sampling rate	Synthetic aperture time	Ground angle	Pulse width	Azimuth resolution
value	200 Hz	20 MHz	24 MHz	150 s	60 deg	500 us	20 m

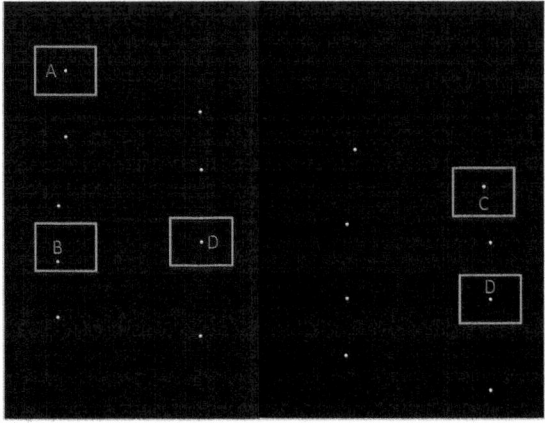

Fig. 3. Lattice imaging results

Table 2. GEOSAR evaluation results of each point of ground angle

Target		A	B	C	D	E
Distance direction	PSLR(dB)	−13.96	−13.32	−13.33	−13.66	−13.19
	ISLR(dB)	−11.25	−10.45	−10.29	−10.91	−10.07
Azimuth	PSLR(dB)	−13.63	−13.27	−13.43	−13.67	−13.26
	ISLR(dB)	−11.14	−10.99	−10.71	−11.45	−10.29

5 Conclusion

A frequency domain imaging processing algorithm suitable for GEO SAR is proposed in this paper. Firstly, the two-dimensional parameters of the original echo are analyzed, then the echo domain is compensated in time domain, and then the two-dimensional NCS imaging is processed. The algorithm can also achieve large skew angle large scene imaging processing. The simulation results show that the algorithm is scientific and applicable and provides a basis for further theoretical research.

References

1. Cheney M. A mathematical tutorial on synthetic aperture radar[J]. SIAM review. 2001, 43 (2):301–312.
2. Jordan R L, Huneycutt B L, Werner M. The SIR-C/X-SAR Synthetic Aperture Radar System [J]Proc. IEEE, 1991, 79(6):827–838.
3. Kiyo Tomiyasu, Jean L. Pacelli, Synthetic Aperture Radar Imaging from an Inclined Geosynchronous Orbit[J], Geoscience and Remote Sensing, IEEE Transactions on, 1983, GE-21:324–329.
4. Bruno D, Hobbs S E, Ottavianelli G. Geosynchronous synthetic aperture radar: concept design, properties and possible applications[J]. Acta Astronautica. 2006, 59(1): 149–156.
5. Long T, Dong X, Hu C, et al. A new method of Zero-Doppler centroid control in GEO SAR [J]. IEEE Geosci. Remote Sens. Lett. 2011, 8(3): 513–516.

Sea–Land Segmentation Algorithm for SAR Images Based on Superpixel Merging

Shiyuan Chen[✉] and Xiaojiang Li

Department of Aerospace Science and Technology, Space Engineering University, Beijing 101416, China
anmyhlydlc@163.com

Abstract. Aimed at improving the accuracy of sea–land segmentation for synthetic aperture radar (SAR) images, a novel sea–land segmentation approach based on superpixel merging is proposed. First, pixel dissimilarity measure is defined on the basis of simple linear iterative cluster (SLIC) algorithm and image presegmentation steps are given to generate superpixels to prepare primitive subregions for the following merging process. Second, three rules, including vicinity rule, edge rule, and similarity rule are proposed to guide the superpixel merging and then a coarse-fine merging strategy is presented to complete sea–land segmentation. Experimental results show that the proposed method has higher segmentation accuracy for SAR images compared with other algorithms.

Keywords: Synthetic aperture radar (SAR) · Sea–land segmentation · Superpixel merging · Coarse merging · Fine merging

1 Introduction

Sea–land segmentation is one of the key steps to implement the ship target detection using synthetic aperture radar (SAR) imagery, and the segmentation accuracy has an important influence on coastal target detection. What's more, the extracted coastline can be applied many activities such as automated navigation, geographic mapping, coastal environmental protection, etc. [1]. Therefore, SAR image sea–land segmentation has received the researchers' extensive concern. However, the contrast between sea and land is weak and the boundary is not clear, due to speckle and strong ocean currents, which brings a big challenge for sea–land segmentation with high accuracy.

In recent decades, many approaches have been proposed for sea–land segmentation, such as edge detection [2], clustering [3], thresholding [4–6], etc. Those algorithms above are simple and intuitive, and have good operability. But they are easily affected by noise and the extracted coastline is always discontinuous. Subsequently, Markov random field (MRF) and level set method are presented to improve the segmentation accuracy, whose large computational burden makes them less real-time [7]. With the development of SAR imaging technology, SAR images obtained always have

© Springer Nature Singapore Pte Ltd. 2019
L. Wang et al. (eds.), *Proceedings of the 5th China High Resolution Earth Observation Conference (CHREOC 2018)*, Lecture Notes in Electrical Engineering 552,
https://doi.org/10.1007/978-981-13-6553-9_3

a high resolution and large geographical coverage, making the content more complex and bringing more severe challenges to segmentation. To deal with the problems above, low-level and high-level image segmentation algorithms are combined together to get the coastline [8]. Zhang et al. [9] propose a novel object-based region-growing and edge detection algorithm, which use a multiresolution method to generate the object units for region growing. Ning et al. [10] use mean shift to generate subregion for merging and obtains the final segmentation results according to the maximum similarity between regions. However, it takes so much time when those methods are utilized to process wide-swath SAR images. Thus, Liu et al. [8] propose a region-merging approach for coastline extraction. It first extracts image feature and uses modified K-means method for presegmentation. Then a coarse-fine strategy is presented to get the final coastline. The performances on Sentinel-1A SAR data demonstrate the algorithm's accuracy and effectiveness.

The methods mentioned above perform in the similar way to get the final sea–land segmentation results. The SAR image is first presegmented to obtain the subregions for the following merging phrase, and then different region-merging strategies are utilized to obtain the final results. It is derived that image presegmentation can affect the accuracy and an efficient merging strategy can not only help to get a good result but also low the computational burden. In order to improve the accuracy of sea–land segmentation, a novel sea–land segmentation approach based on superpixel merging is proposed. First, a modified simple linear iterative cluster (SLIC) is designed to generate superpixel for merging. Then, three rules for merging are introduced and a coarse-fine merging strategy is presented to complete sea–land segmentation.

The rest of the paper is organized as follows. Section 2 describes the method to generate superpixels. Then the rules and a strategy for superpixel merging are given in Sect. 3. Experimental results and conclusions are provided in Sects. 4 and 5.

2   Superpixel Generation

Traditional segmentation methods usually take a single pixel as the basic unit for merging, resulting in large computational complexity and bringing some difficulties to the practical application because the wide-swath SAR images have so many pixels. To solve the problem, images are usually presegmented before merging and several sub-regions are produced. The SLIC algorithm is recently presented superpixel segmentation algorithm that shows good performance in superpixel generation for optical images [11]. Thus, the SLIC algorithm is first introduced in the paper and then the improved strategy is proposed and applied to SAR image presegmentation to generate superpixels.

2.1   SLIC Algorithm

The SLIC algorithm first converts image from RGB color space to CIE-Lab color space and obtains 5-D $[l, a, b, x, y]$ space, where $[x, y]$ is the pixel position and $[l, a, b]$ is the pixel color. Then compute pixel dissimilarity and adopt k-means clustering to generate superpixels. The pixel dissimilarity is calculated as follows.

Let $[l_i, a_i, b_i, x_i, y_i]$ and $[l_j, a_j, b_j, x_j, y_j]$ represent pixels i and pixel j 5-D space, the CEI-Lab color dissimilarity d_p can then be defined as:

$$d_p = \sqrt{\left(l_i - l_j\right)^2 + \left(a_i - a_j\right)^2 + \left(b_i - b_j\right)^2} \tag{1}$$

The spatial distance d_s is defined as:

$$d_s = \sqrt{\left(x_i - x_j\right)^2 + \left(y_i - y_j\right)^2} \tag{2}$$

Finally, the dissimilarity measure $D(i,j)$ of pixels i and j is defined as:

$$D(i,j) = \sqrt{\left(\frac{d_p}{m}\right)^2 + \left(\frac{d_s}{S}\right)^2} \tag{3}$$

where m is a parameter controlling the relative weight between d_p and d_s, $S = \sqrt{N/K}$, N is the number of pixels and K is the number of superpixels. Larger $D(i,j)$ represents bigger difference between pixels i and j.

2.2 Presegmentation Based on Modified SLIC

It is known that pixels in the single-channel SAR images only have intensity information and traditional SLIC cannot be directly applied to image segmentation. Thus, a new pixel dissimilarity measurement is defined in this paper and a modified SLIC is introduced for image presegmentation.

2.2.1 Dissimilarity Measure

The usual way to measure the dissimilarity of pixels i and j is to simply compare their intensities, which is not effective because of the speckles. To solve the problem, the intensities of the two pixels and the two local patches centering these two pixels are utilized to measure the dissimilarity. The intensity dissimilarity d'_p of pixel i and j is defined as:

$$d'_p = \sqrt{\left(I_i - I_j\right)^2 + \left(\bar{I}_{u_i} - \bar{I}_{u_j}\right)^2} \tag{4}$$

where u_i and u_j are the patches entering pixels i and j separately, and $\bar{I}_{u_i}$ and $\bar{I}_{u_j}$ represent the average intensities in u_i and u_j.

The mean value of the pixel patch is used to represent the value of the central pixel, avoiding the interference caused by the too large or too small brightness of a single pixel. And in order to make each patch describe the central pixel better, the patch size should not be too large and experiments show that a 5×5 patch is found to be appropriate.

The pixel spatial distance is still expressed as a European distance as shown in (2). Finally, the dissimilarity measure $D'(i,j)$ of pixels i and j can be defined as:

$$D'(i,j) = \sqrt{d'_p + \lambda \cdot d_s} \tag{5}$$

where a parameter λ is used to balance the relative importance of the intensity dissimilarity and the spatial dissimilarity. Experiments show that when λ is in the range [2, 10], it can get accuracy dissimilarity measure. So $\lambda = 2$ is used in this paper.

2.2.2 Image Presegmentation

After pixel dissimilarity definition, the k-means algorithm is adopted to generate superpixels to complete the SAR image presegmentation, which is divided into four steps:

Step 1: The number of the superpixels is set as K, and initialize the cluster centers $C_k, k = 1, 2, \ldots, K$, by sampling pixels at regular grid steps S. To avoid centering a superpixel on an edge or a noisy pixel, move C_k to the smallest gradient position in a 3×3 neighborhood. Set the label $l(i) = -1$ for each pixel i and set dissimilarity $d(i) = \infty$ for pixel i.

Step 2: For each cluster center $C_k, k = 1, 2, \ldots$, the dissimilarity D' between pixel i and cluster center C_k is calculated according to Eq. (5) in a $2S \times 2S$ neighborhood. If $D'(i) < d(i)$, then set dissimilarity $d(i) = D'$, label $l(i) = k$ until all cluster centers are processed.

Step 3: Update each cluster center to be the mean coordinates of the pixels belonging to that cluster center.

Step 4: Repeat Step 2 and Step 3 until the position of cluster center is stable and output clustering results, that is, the generated superpixels.

There are two parameters that affect the effect of image presegmentation: the number of superpixels and compactness coefficient, which can be determined by the experiments.

3 Proposed Segmentation Method

3.1 Superpixel Merging Rules

3.1.1 Vicinity Rule

Vicinity rule is used to determine whether two superpixels are adjacent in space. For each pair of superpixels is expressed as (s_i, s_j), $i \neq j$, $i, j = 1, 2, \ldots, N_s$, where N_s is the number of superpixels, the vicinity is defined as:

$$C_1(i,j) = \begin{cases} 1, s_i \text{ and } s_j \text{ are neighbors} \\ 0, s_i \text{ and } s_j \text{ are not neighbors} \end{cases} \tag{6}$$

$C_1(i,j)$ represents the relative spatial position of a pair of superpixels. If $C_1(i,j) = 1$, then s_i and s_j are of spatial vicinity and satisfy the rule of vicinity and vice versa.

3.1.2　Edge Rule

Edge rule is used to determine whether a superpixel is at the edge of an image. An edge detector with different orientations [12] is used to detect image edge in the paper. Figure 1 shows the structure of the edge detector and the detector is determined by parameters $K_f = \{l, w, d, \theta_f\}$, where l is the length, w is the width, d is the distance between the two rectangles, and θ_f is the angular increment between neighboring orientations. For a detector in the orientation θ_f, first calculate the average intensities $\bar{R}_1(x, y, \theta_f)$ and $\bar{R}_2(x, y, \theta_f)$ of the rectangles on both sides of the center pixel (x, y), then compute the ratio edge strength map $r(x, y, \theta_f)$:

$$r(x, y, \theta_f) = \min\left(\frac{\bar{R}_1(x, y, \theta_f)}{\bar{R}_2(x, y, \theta_f)}, \frac{\bar{R}_2(x, y, \theta_f)}{\bar{R}_1(x, y, \theta_f)}\right) \tag{7}$$

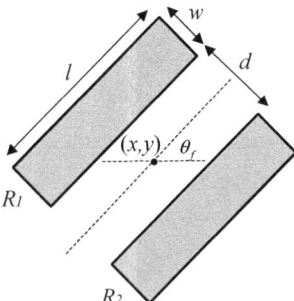

Fig. 1. The structure of edge detector with different orientations

By adjusting the different values of θ_f, $r(x, y, \theta_f)$ with different orientations can be obtained.

The edge strength of pixel (x, y) is denoted as:

$$g(x, y) = 1 - \prod_{f=1}^{M} r(x, y, \theta_f) \tag{8}$$

where M is the number of orientations of the edge detectors.

For pixels in image edges or around them, the ratio edge strength map $r(x, y, \theta_f)$ is anisotropic. The closer it gets to the direction of the edge, the smaller value of it is, while the closer it gets to the vertical direction of the edge, the larger value of it is. Therefore, $g(x, y)$ of the pixel (x, y) at the edge of an image usually gets bigger and is close to 1. Meantime, $g(x, y)$ of the pixel (x, y) inside an image usually gets smaller and is close to 0. The edge strength $g(x, y)$ enhances the edge strength map by using the anisotropic pixels ad the edge of the image and inhibits the edge strength map by using the isotropic pixels inside the image. Thus, whether a pixel is located at the image edge can be determined according to the value of the edge strength $g(x, y)$.

So, for any superpixel $s_i, i = 1, 2, \ldots, N_s$, its edge strength $C_2(i)$ is expressed by:

$$C_2(i) = \frac{\sum_{(x,y)\in s_i} g(x,y)}{s_i} \tag{9}$$

3.1.3 Similarity Rule

Similarity rule is used to measure the similarity of two superpixels. The feature vector of the superpixel is first extracted and then vector degree of match (VDM) criterion [13] is adopted to represent the similarity of two feature vectors in order to measure the similarity of two superpixels.

For any superpixel s_i, $i = 1, 2, \ldots, N_s$, its feature vector $F(i)$ can be obtained as follows:

Step 1: Compute the feature vector $F(x, y)$ of any pixel $(x, y) \in s_i$;
Step 2: Calculate the feature vector $F(i)$ of superpixel s_i, $i = 1, 2, \ldots, N_s$ by the average of the feature vectors of pixel belonging to s_i:

$$F(i) = \frac{\sum_{(x,y)\in s_i} F(x,y)}{|s_i|} \tag{10}$$

For step 1, a fusion strategy that combines intensity feature and texture feature is proposed to extracted feature vector $F(x, y)$ in this paper:

$$F(x,y) = [w_1, w_2] \times \begin{bmatrix} F_1(x,y) \\ F_2(x,y) \end{bmatrix} \tag{11}$$

where $F_1(x, y)$ is the intensity feature of pixel (x, y), $F_2(x, y)$ is the texture feature of pixel (x, y). w_1 and w_2 are the corresponding weights, satisfying $w_1 + w_2 = 1$.

The pixel intensity feature $F_1(x, y)$ can be obtained by normalizing the intensity image to $[0, 1]$. The Gabor filter with different orientations and scales can describe the image texture accurately [14] and is utilized to obtain texture feature $F_2(x, y)$. The input image is filtered by a bank of Gabor filters to get the Gabor coefficient vectors and then, the coefficient vectors are normalized to $[0, 1]$ to obtain texture feature. Set center frequency of a Gabor filter as $\omega_k (k = 1, 2, \ldots, N_\omega)$, orientation as $\theta_l (l = 1, 2, \ldots, N_\theta)$, then the weights of the Gabor coefficient vectors satisfy:

$$\sum_{k=1}^{N_\omega} \sum_{l=1}^{N_l} w_2(\omega_k, \theta_l) = w_2 \tag{12}$$

The weights of the Gabor coefficient vectors are set as equal weights, then

$$w_2(\omega_k, \theta_l) = \frac{w_2}{N_\omega \times N_\theta} \tag{13}$$

After the calculation of the feature vector of each pixel, the feature vector $F(i)$ of superpixel s_i, $i = 1, 2, \ldots, N_s$ can be obtained according to formula (10).

For a pair of superpixels (s_i, s_j), $i \neq j$, $i,j = 1, 2, \ldots, N_s$, the feature vector is $F(i)$ and $F(j)$ separately, energy difference is σ_1, angular difference is σ_2, then σ_1 can be defined as

$$\sigma_1(F_i, F_j) = \min \left\{ \frac{|F_i|}{|F_j|}, \frac{|F_j|}{|F_i|} \right\} \tag{14}$$

where $|F_i|$ and $|F_j|$ are the moduli of F_i and F_j, respectively.

$$\sigma_2(F_i, F_j) = \frac{(\pi - \alpha)}{\pi} \tag{15}$$

where $\alpha = \arccos(F_i \circ F_j / (|F_i| \cdot |F_j|))$ and $F_i \circ F_j$ is the scalar product of F_i and F_j.

Sum up, the VDM value σ can be denoted as:

$$\sigma(F_i, F_j) = \sigma_1(F_i, F_j) \cdot \sigma_2(F_i, F_j) \tag{16}$$

σ is then normalized to [0, 1]. Moreover, the bigger the value of σ is, the more similar F_i and F_j are.

3.2 Coarse to Fine Strategy for Superpixel Merging

According to the visual features of the human eye, when people see an image, they first notice different objects and then consider the details between objects like edges, shape, and so on. Therefore, a coarse-fine superpixel merging strategy is presented which consists of two stages: coarse merging stage (CMS) and fine merging stage (FMS). After image presegmentation, the superpixels can be roughly divided into two categories. One is those within the ocean area or the land area, whose category can be clearly determined. The other is the superpixels that are located between the sea and land, which are doubtful to be merged. The object of CMS is to merge the superpixels without ambiguity while the object of FMS is to merge the superpixels with ambiguity. The details of CMS and FMS are introduced separately.

3.2.1 Coarse Merging Stage (CMS)

As for implementation, CMS merges superpixels based on the vicinity rule and the edge rule. CMS first finds all adjacent pairs of superpixels D_1 based on the vicinity rule:

$$D_1 = \{(s_i, s_j) | C_1(i, j) = 1, i \neq j, i, j = 1, 2, \cdots, N_s\} \tag{17}$$

Then, CMS chooses the pair of superpixels D_2 satisfying the merging condition as follows based on the edge rule:

$$D_2 = \{(s_i, s_j) | C_2(i) \leq \beta, C_2(j) \leq \beta, (s_i, s_j) \in D_1\} \tag{18}$$

where β is a parameter controlling the ratio of the number of superpixels in CMS to the total number of superpixels. If β is too large, most of the superpixels are merged in CMS and the accuracy of sea–land segmentaion will be affected. If β is too small, those superpixels without ambiguity need to be merged in FMS, leading to a high computational burden. Experiments show that when β is in the range of [4/5, 9/10], the algorithm can obtain a better segmentation result.

Before merging, the "sea" seed and "land" seed should be selected respectively from the superpixels as the initial merging superpixels. The "sea" seed can be denoted as:

$$\text{Seed}_{\text{sea}} = \arg \min_{s_i} \left(\frac{\sum\limits_{(x,y) \in s_i} I(x, y)}{|s_i|}, C_2(i) \leq \beta \right) \tag{19}$$

The "land" seed can be denoted as:

$$\text{Seed}_{\text{land}} = \arg \max_{s_i} \left(\frac{\sum\limits_{(x,y) \in s_i} I(x, y)}{|s_i|}, C_2(i) \leq \beta \right) \tag{20}$$

The steps of CMS are as follows:

Step 1: Select the "sea" seed and "land" seed as the initial merging superpixels;
Step 2: Find all the adjacent pairs of superpixels D_1 satisfying the vicinity rule based on Eq. (17);
Step 3: Choose the pair of superpixels D_2 satisfying Eq. (18) and merge them until there are no such pair of superpixels existed.>

3.2.2   Fine Merging Stage (FMS)

After CMS, the merged sea region is denoted as R_{sea} and the merged land region is denoted as R_{land}. And the unmerged superpixels are located at the junction of the sea and land. So, it is difficult to judge those categories by brightness value alone. As for implementation, FMS merges superpixels based on the similarity rule.

The unmerged superpixels are denoted as s_i, $i = 1, 2, \ldots, N_u$, N_u, and N_u is the number of unmerged superpixels.

The steps of FMS are as follows:

Step 1: Find the superpixels belonging to sea and land separately around the unmerged superpixel s_i, that is $s_s \in R_{\text{sea}}$, $s_l \in R_{\text{sea}}$;
Step 2: Compute the feature vector F_i of s_i, the feature vector F_s of s_s and the feature vector F_l of s_l according to Eq. (10);
Step 3: Calculate $\sigma(F_i, F_s)$ and $\sigma(F_i, F_l)$ separately based on Eq. (16) and compare their values. If $\sigma(F_i, F_s) > \sigma(F_i, F_l)$, the superpixel s_i is merged into sea region and vice versa;

Step 4: Determine the other unmerged superpixels' categories and merge them until all the superpixels have been processed.

3.3 Flowchart of the Proposed Algorithm

Combined with the generation of superpixels using modified SLIC, a sea–land algorithm based on superpixel merging is designed and the flowchart and the detailed flowchart of it is shown in Fig. 2, which can be divided into five steps:

Step 1: Input SAR image and perform image presegmentation to generate superpixels using modified SLIC;
Step 2: As for CMS, merging the superpixels based on the vicinity rule and the edge rule;
Step 3: As for FMS, merging the superpixels based on the similarity rule;
Step 4: Output the sea–land segmentation result.

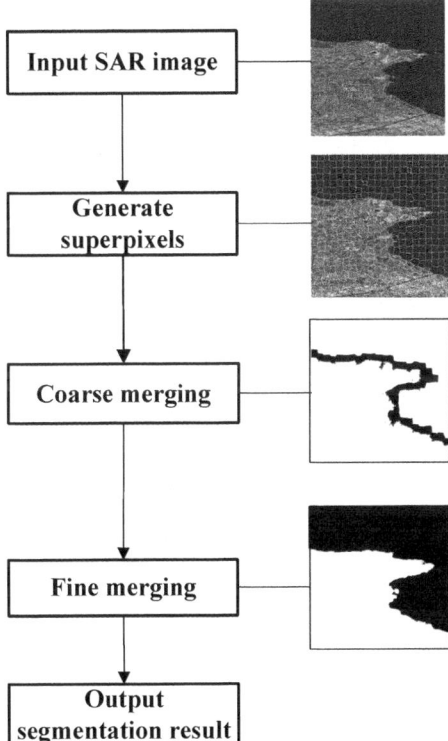

Fig. 2. The flowchart of sea–land segmentation algorithm based on superpixel-merging

4 Experimental Results and Discussion

To demonstrate the effectiveness of the proposed approach, a typical Sentinel-1A product, with a resolution of 5 m × 20 m and a size of 1300 × 1300 pixels, acquired in Interferometric Wide (IW) Swath, is exploited. Moreover, to better elevate this approach, it is compared with the two-dimensional Otsu (2d-Otsu) method [4] and the MKAORM [8].

When conducting the experiments, the parameters K and C should be determined first and other parameters are set as follows: for the edge detectors, the number of orientations is set as $M = 8$ and for the Gabor filter bank, four orientations and three scales are adopted. In this paper, the parameters K and C are adjusted to determine the values which can help achieve the best presegmentation effect. Figure 3 shows the superpixel segmentation results. From Fig. 3b–d, it is observed that if the number of superpixels K is too small, some sea regions and land regions will be merged into the same superpixel, while if the number of superpixels K is too large, the computational burden will be too high. From Fig. 3d–f, it can be seen that if the compactness C is too small, the shape of superpixels will be extremely irregular, while if the compactness C is too large, some image details will be incorrectly segmented. In conclusion, $K = 500$, $C = 100$ is used to get a better presegmentation result.

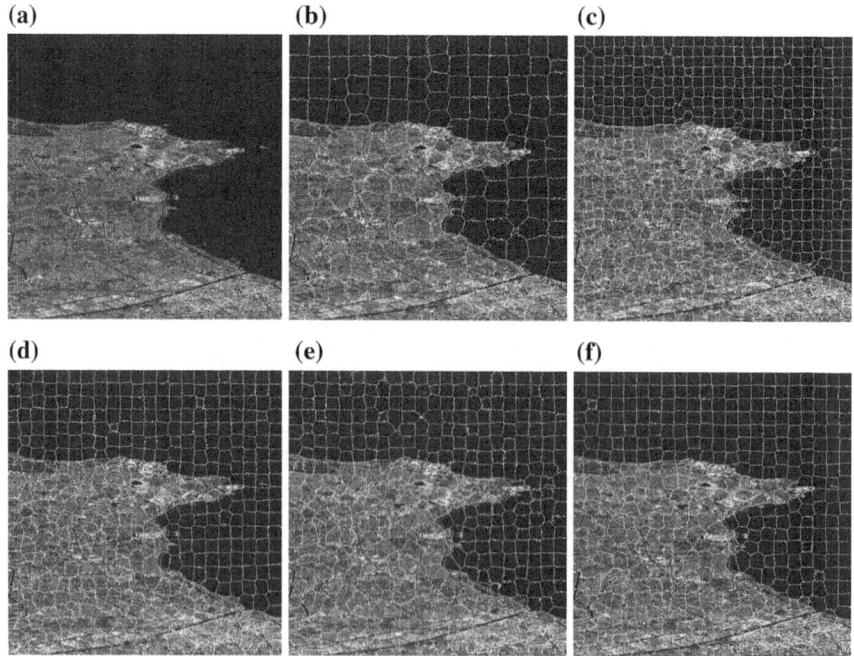

Fig. 3. The superpixel results of different parameters. **a** Original SAR image; **b** $K = 200$, $C = 100$; **c** $K = 800$, $C = 100$; **d** $K = 500$, $C = 100$; **e** $K = 500$, $C = 80$; **f** $K = 500$, $C = 120$

(a)   (b)   (c)   (d)

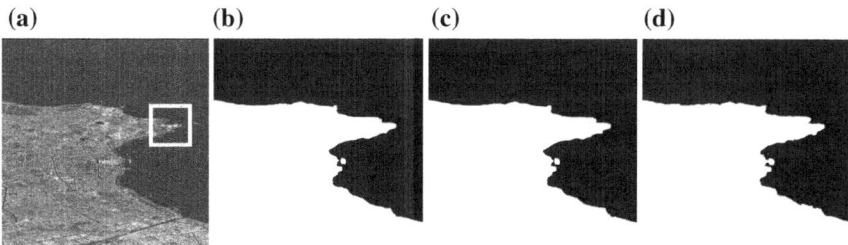

Fig. 4. Sea–land segmentation results for different algorithms. **a** Original SAR image; **b–d** are the segmentation results produced by 2d-Otsu, MKAORM and our proposed algorithm, respectively

The sea–land segmentation results of the three algorithms are shown in Fig. 4. In the regions where the contrast between sea and land is high, all three algorithms can get good segmentation results. To better observe the segmentation in the areas with low contrast, the white box area in Fig. 4a is enlarged and the segmentation results are shown in Fig. 5. It is observed from Fig. 5 that in the area with low contrast between ocean and land, both the 2d-Otsu method and MKAORM method identify the ocean as land by mistake, while our proposed approach can segment the ocean and land accurately.

(a)   (b)   (c)   (d)

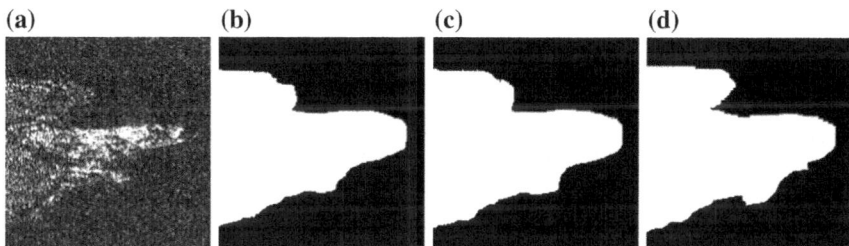

Fig. 5. Sea–land segmentation results of local area for different algorithms. **a** Local area of Original SAR image; **b–d** are the segmentation results of local area produced by 2d-Otsu, MKAORM and our proposed algorithm, respectively

To quantitatively measure the segmentation performance of those three algorithms, the segmentation quality is put forward to measure the segmentation precision [5]. The segmentation quality is defined as:

$$\text{quality} = \frac{N_{dt}}{N_{rt} + N_{fd}} \tag{21}$$

where N_{dt} is the number of extracted sea area pixels, N_{rt} is the number of true pixels, and N_{fd} is the number of extracted false pixels. The segmentation quality considers both the false alarm rate and missing alarm rate. And the closer its value is to 1, the better segmentation effect is.

Table 1. Comparison of the segmentation quality

Segmentation algorithms	2d-Otsu	MKAORM	The proposed
Quality	95.79	96.53	98.72

Based on the sea–land segmentation results produced by artificial demarcation, the segmentation quality of three algorithms can be computed, which is shown in Table 1. It can be seen that our proposed algorithm has a higher segmentation precision than other two algorithms. In conclusion, the proposed approach is more applicable to segment SAR images which have low contrast between sea and land.

5 Conclusions

The paper presents a new sea–land segmentation algorithm for SAR imagery, which combines the modified SLIC algorithm and a coarse-fine merging strategy. The traditional SLIC algorithm is first modified to presegment the SAR image to prepare superpixels for the merging phrases. Furthermore, three rules which are utilized to instruct how to merge superpixels is designed and a coarse-fine merging strategy is presented to fulfill the segmentation. Experiments show that the proposed algorithm can have a good performance on Sentinel-1A IW mode imagery with large spatial coverage.

References

1. Huang Kuihua, ZHANG Jun. A coastline detection method using SAR images based on the local statistical active contour model [J]. Journal of Remote Sensing, 2011, 15(4):737–749.
2. Liu H, Jezek K C. Automated extraction of coastline from satellite imagery by integrating Canny edge detection and locally adaptive thresholding methods [J]. International Journal of Remote Sensing, 2004, 25(5):937–958.
3. Liu Y. Coastline detection from remote sensing image based on K-mean cluster and distance transform algorithm [J]. Advanced Materials Research, 2013, 760–762:1567–1571.
4. Liu Jianzhuang, Li Wenqing. The Automatic Thresholding of Gray-Level Pictures via Two-Dimensional OTSU Method [J]. Acta Automatica Sinica, 1993, 19(1):101–105.
5. An Chengjin, Niu Zhaodong, Li Zhijun, et al. Otsu Threshold Comparison and SAR Water Segmentation Result Analysis [J]. Journal of Electronics and Information Technology, 2010, 32(9):2215–2219.
6. Song Wenqing, Wang Yinghua, Lu Hongxi, et al. Otsu segmentation algorithm for SAR images based on power transformation [J]. Systems Engineering and Electronics, 2015, 37 (7):1504–1511.

7. Chen Xiang, Sun Jun, Yin Kuiying, et al. Sea-land Segmentation Algorithm of SAR Image Based on Otsu Method and Statistical Characteristic of Sea Area [J]. Journal of Data Acquisition and Processing, 2014, 29(4):603–608.

8. Z Liu, F Li, N Li, et al. A Novel Region-Merging Approach for Coastline Extraction From Sentinel-1A IW Mode SAR Imagery [J]. IEEE Geoscience and Remote Sensing Letters, 2016, 13(3):324–328.

9. T Zhang, X Yang, S Hu, et al. Extraction of coastline in aquaculture coast from multispectral remote sensing images: Object-based region growing integrating edge detection [J]. Remote Sensing, 2013, 5(9):4470–4487.

10. J Ning, L Zhang, D Zhang, et al. Interactive image segmentation by maximal similarity based region merging [J]. Pattern Recognition, 2010, 43(2):445–456.

11. R Achanta, A Shaji, K Smith, et al. SLIC superpixel compared to state-of-the-art superpixel methods [J]. IEEE Transactions on Pattern Analysis and Machine Intelligence, 2012, 34 (11):2274–2282.

12. Schou J, Skriver H, Nielsen A A, et al. CFAR edge detector for polarimetric SAR images[J]. IEEE Transaction on Geoscience and Remote Sensing, 2003, 41(1):20–32.

13. A Baraldi, F Parmiggian. Single linkage region growing algorithms based on the vector degree of match [J]. IEEE Transaction on Geoscience and Remote Sensing, 1996, 34 (1):137–148.

14. Zeng Shuyan, Zhang Guangjun, Li Xiuzhi. Image target distinguish based on Gabor filters [J]. Journal of Beijing University of Aeronautics and Astronautics, 2006, 32(8): 954–957.

Vector Quantization: Timeline-Based Location Data Extraction and Route Fitting for Crowdsourcing

Naiting Xu[1,2(✉)], Yi Wang[1,2], Xing Chen[1,2], and Haiming Lian[1,2]

[1] Institute of Electronics, Chinese Academy of Sciences, Suzhou 215123, Jiangsu, China
sa615237@mail.ustc.edu.cn
[2] Key Laboratory of Intelligent Aerospace Big Data Application Technology, Suzhou 215123, Jiangsu, China

Abstract. Because of the superiority of crowdsourcing, many companies are eager to seek the solutions for some issues through crowdsourcing, e.g., assembling crowd wisdom crystallization for a wonderful design, collecting personal-privacy information or geographical location information, etc. Location services are becoming more and more important to the lives and lifestyles of modern people, so the Internet companies and traditional enterprises placed great emphasis on this theme. We can obtain valuable information from the large-scale datasets, but in certain circumstances, a large amount of data will be a burden for us to analyze problems. On account of redundant data that carries little weight with us, in this paper, we mainly consider trajectory data extraction of timeline-based location information and route fitting in the field of military (e.g., GIS) and civilian (e.g., traffic). For the purpose, we refer to the idea of Vector Quantization, denoted as VQ, and give a model of TDERTM (Timeline-based Data Extraction and Route Fitting Model) to solve that. In this sense, it is efficient and convenient to determine whether the target (ship, aircraft, etc.) cross the border and draw a navigation trajectory with low redundancy.

Keywords: Crowdsourcing · Data extraction · Route fitting · VQ · Navigation trajectory

1 Introduction

Crowdsourcing connects unobtrusive and ubiquitous sensing technologies to create solutions that improve the convenience of our research work. Over the past decade, we have witnessed the rise and development of crowdsourcing, as well as its great application value in the field of navigation, delivery services and cooperative search, etc. Crowdsourcing services have emerged in many well-known Internet companies. For example, requesters can outsource various tasks via the Amazon Mechanical Turk [1]. For example, the Google Helpouts [2], an online collaboration website, encourages people to provide online help. Typical examples include Yahoo! Answers [3] (a question-and-answer platform), Wikipedia [4] (encyclopedia collaborative program), Coined in 2005 (a promising production model). Online services such as these have

© Springer Nature Singapore Pte Ltd. 2019
L. Wang et al. (eds.), *Proceedings of the 5th China High Resolution Earth Observation Conference (CHREOC 2018)*, Lecture Notes in Electrical Engineering 552,
https://doi.org/10.1007/978-981-13-6553-9_4

initiated huge transformation in human society. Here, what we are trying to do now is collect timeline-based navigation trajectory information data for researchers' work in civilian and military fields. Trajectories are spatiotemporal data, comprising the geographical positions of moving objects in various time steps. Moving objects can be vehicles in roads, airplanes in airline traffic, animals in forest areas, and so on. Capturing and analyzing this type of data may have numerous potential applications.

Typically, units involved in crowdsourcing will send location information to crowdsourcing platforms by mobile devices such as smartphones, compasses and other devices that can send signals. In geographical information system, and technical personnel to analyze the sea area ship, aircraft navigation behavior, whether the border, is a violation of the relevant laws and regulations, we all need to track on the basis of judgment, and the effective way to avoid such incidents happen repeatedly. Then, if you simply connect the huge Beidou point information on time, it will make the track messy and bring some burden to the technical personnel analysis work. Therefore, how to filter out redundant Beidou point information is the problem to be solved in this paper. In addition, in the civil aspects, the performance of the Austrian technology is quite outstanding. Extreme Austrian technology has established a massive, high frequency vehicle data collection platform, collecting vehicle track data, vehicle sensor data, and vehicle rear image data in crowdsourcing mode. The production and distribution of autopilot maps are more than 90% automated. If we want to study people's timeline-based travel trajectory in a region, we can summarize some people's behavior habits and provide more convenient services for people. In terms of intelligent transportation, it will be a great progress. However, travel behavior is unpredictable. For example, when people travel, they stay in a certain area for a long time. From the point of view of real-time data collection, this area will produce a lot of redundant information points. So if all locus points in this area are considered, the track fitting based on time will be confused, which is not conducive to the summary of human behavior. What we need is clear lines that can eliminate redundancy points as far as possible, and clear routes that are very useful for our research and work.

2 Related Work

Xie Jianhua et al. proposed an algorithm for analyzing simulated trajectory data based on density clustering [5]. Using the large-scale data provided by GPS to analyze, it provides a good reference value in the field of agricultural machinery operation. Zahedeh Izakian et al. proposed an automatic clustering technique for trajectory data based on particle swarm optimization (PSO) [6]. The dynamic time warping (DTW) distance is considered as one of the most commonly used distance measures for trajectory data. For example, experts in traffic control and urban planning are interested in the structure available in the vehicle movement mode (such as cars, buses, aircraft) at different intervals. They can use this information for road construction or design monitoring systems and so on.

3   Contribution and Paper Structure

3.1   Summary of Key Contributions

The following is a list of our main contributions.

- The purpose of this paper is to judge the state of a large amount of position information through a few position point sets and to improve the efficiency of analysis. For example, to determine whether a ship is staying in a certain area for a long time, or if we want to know a person's behavior, it takes only a few points to analyze the result.
- In this paper, two fitting methods are put forward, and their advantages and disadvantages are compared and analyzed. One solution uses density clustering technology to remove redundant point sets, and the other is to draw lessons from the idea of vector, adopt the method of drawing grid, and make K-means clustering for point sets in each grid.
- Two methods of line fitting are compared and analyzed with experiments.

3.2   Paper Structure

The remainder of this paper is organized as follows. In Sect. 2, we review the state-of-the-art research on the application of clustering technology. In Sect. 4, we summarize the preliminaries about crowdsourcing system and line fitting model. In Sect. 5, we show related algorithms. In Sect. 6, the experiment and theoretical analysis about line fitting models are given. In Sect. 7, we draw our conclusions.

4   Preliminaries

In this section, we will summarize system overview, related technology, line fitting method based on clustering, and line fitting method based on grid. Table 1 lists frequently used notations.

Table 1. Frequent used notations

Notation	Description
V	Travel speed
E	The density radius of clustering
MinPts	Clustering density
L	Cell mesh size
cVector	Track points after data extraction

4.1 System Overview

The purpose of this paper is to design an analyzer which can extract the representative position information and solve the problem of weak adaptive ability in data extraction and line fitting. The design can extract the typical data sets and fit the trace lines containing less data. Generally speaking, the number of clusters or the density of clusters must be preset first in the clustering method, so setting appropriate clustering parameters is the key factor to determine the clustering results. In most cases the trajectory is erratic and the reasonable clustering parameters can be calculated automatically according to the specific conditions so the adaptability of the line fitting can be extended better. In addition, the line fitting can avoid the unstable clustering result caused by the unreasonable setting of clustering parameters.

The workflow of this system is as follows (Fig. 1).

- Figure 1 shows the process of data acquisition and analysis. Each unit or individual involved in crowdsourcing must be equipped with a corresponding point transmitting device, and set the corresponding transmitting frequency f, which can be set according to different traffic equipment. Each unit sends the information with a category number. After the task, the platform will classify the information according to the category number.
- Then the redundant information is filtered out by the algorithm model, and the remaining information is connected according to the time to form a representative trajectory line. The detailed flow is shown in Fig. 2.

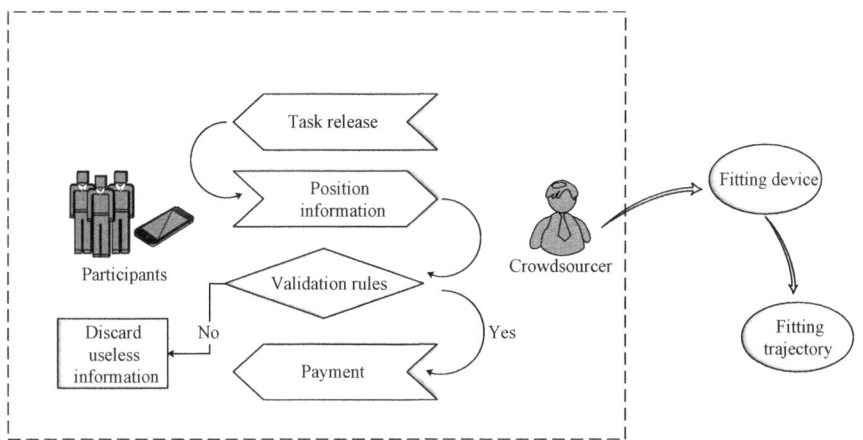

Fig. 1. Crowding system

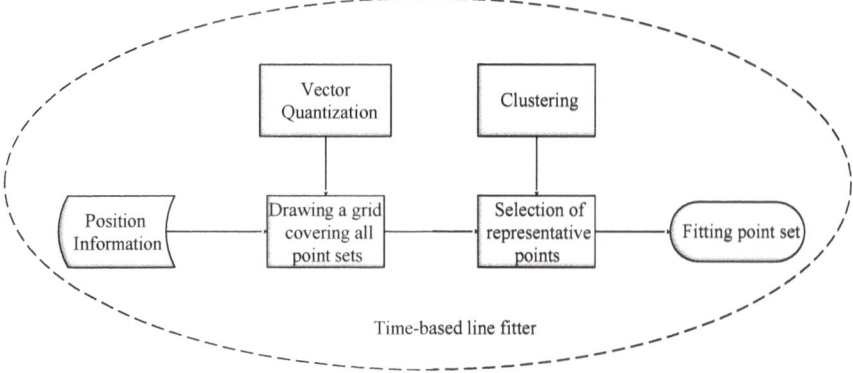

Fig. 2. Time-based line fitter

Fig. 3. VQ application example

4.2   Techniques

The technical scheme adopted by the invention is that the information of position data is extracted by using the idea of vector quantization and clustering, and the representative circuit is fitted out.

4.2.1   Vector Quantization

VQ is widely used in many fields [7–11]. Image coding is a good example. For a gray image, the compression scheme of selecting representative points by clustering is undoubtedly a better scheme. The actual practice is to treat each pixel as a data, run K-means, get k centroids, and then use the pixel values of these centroids to replace the pixel values of all points in the corresponding cluster. Color images can also be done in the same way, such as Fig. 3. For example, in RGB trichromatic images, each pixel is treated as a point in a three-dimensional vector space. Similarly, the idea of vector quantization can be applied to track fitting design. For example, the line point set can be simulated as a black pixel in a gray image, so that the representative point set of the line can be extracted by imitating the image compression method.

4.2.2   Cluster

Clustering technology [12–16] has become an important means to analyze big data. The locus point set is clustered, each cluster center point is calculated, and a new locus line is formed according to the point set which is extracted and analyzed according to the time link. Clustering is proposed in two ways. The first method first determines the number of clusters K, then K-mean clustering is applied to all point sets, the other way is to determine the size of the mesh, and then calculate the center point of each grid.

Dean distance is a special form of Minkov distance. Minkowski distance, a measure in Euclidean space, is regarded as a generalization of Euclidean distance, which is a special case of Minkowski distance.

Definitions 1 (Minkowski Distance):

$$\rho(A, B) = \left(\sum_{i=1}^{n} |a_i - b_i|^p \right)^{\frac{1}{p}}$$

In the Minkowski distance formula, the Euclidean distance is the Euclidean distance when $p = 2$, the Manhattan distance when $p = 1$, and the Chebyshev distance when $p = \infty$.

Following the compression method of the photo, we specify the size of the following grid as Mesh Size.

Definitions 2 (Mesh Size):

$$L = \frac{\sum_{i:i \in G, v_i > \max_{j \in G} v_j * 0.5} v_i}{\sum_{i:i \in G, v_i > \max_{j \in G} v_j * 0.5} 1} t$$

v_i is the speed of i th trajectory point. G is a set of all trajectory points, $t = 1$ s.

5   Of Model

5.1   Algorithm of DB-SCAN Model

Algorithm 1: DB-SCAN Algorithm

Input: *E, MinPts*
Output: *cVector*
1: *DBSCAN(E, MinPts) → k clusters*
2: *Calculate the center point of each cluster → k centerpoints*
3: *cVector ← k centerpoints*
4: return *cVector*

Algorithm 1 sets reasonable clustering density and density radius according to the speed of travel and divides all points set into k classes by density clustering method, and calculates the center points of each class. Finally, all the central point sets are synthesized to the line.

5.2   Algorithm of Grid-K-Means Model

Algorithm 2: Grid-K-means Algorithm

Input: *L*
Output: *cVector*
1: *A grid of drawing unit size L covers all point sets*
2: *The point set in each grid is k-means (k = 1) to get the center point of the point set in each grid. k centerpoints*
3: *cVector ← k centerpoints*
4: return *cVector*

According to the travel speed and clustering technique, Algorithm 2 sets a reasonable mesh size L, covers all the locus point sets with the grid, carries out k-means on the points set in each grid, and calculates the cluster center points of each grid. Finally, all the center points are connected into lines according to the track time.

6   Experiment and Theoretical Analysis

In order to show the effect of circuit fitting, we did the following experiments. Table 2 lists the parameters required for the experiments. In Figs. 4, 5 and 6, the horizontal axis is the X-axis and the longitudinal axis is the Y-axis.

Table 2.   Data setting

Setting	Description
I	X,Y: random[0,50]
II	X,Y: incremental random[0,50], random step length [5,15]
III	X,Y: incremental random[0,100], random step length [5,15]

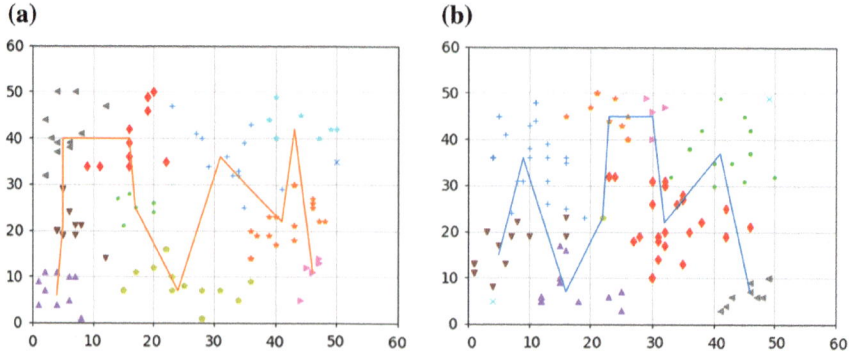

Fig. 4.   a, b DB-Scan model by setting I

Figure 4 shows the DB-SCAN results of setting I point set. The vertex of the broken line is the center of the cluster, and the broken line is the line after the center point is fitted. By observing Fig. 4a, b, it can be found that the method for density clustering can filter out the point set for a certain extent, and then fit a more concise time-based route. The set of points used is dropped, and a more concise time-based route is then fitted. Due to the uniform distribution of the trajectory in a certain region, we can not clearly determine whether the fitting scheme is effective or not. To this end, we take incremental track data to show the effect of route fitting.

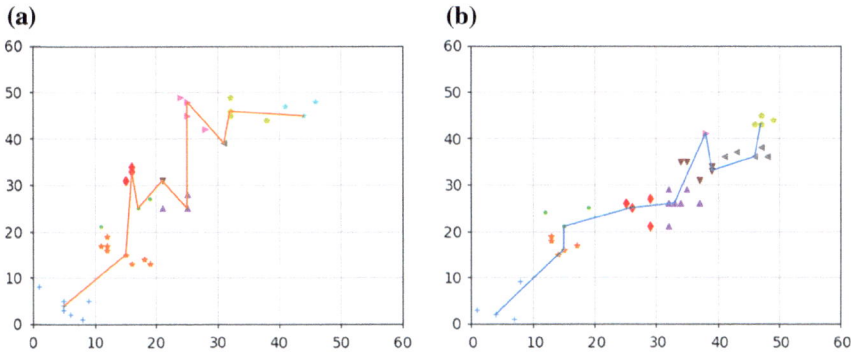

Fig. 5. **a**, **b** VQ model of setting II

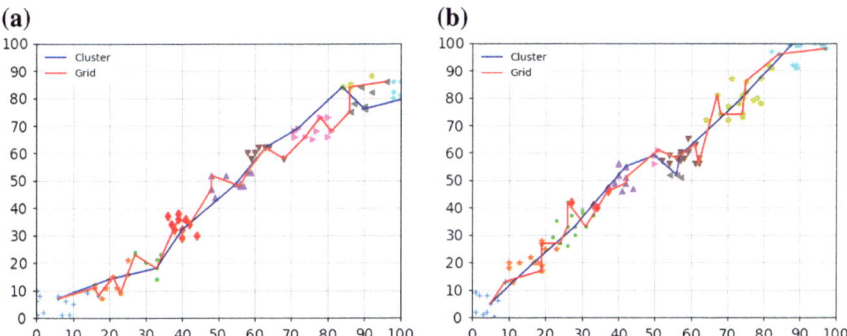

Fig. 6. **a**, **b** Comparison result of setting III

Figure 5 shows the VQ results of setting II point set. The vertex of the broken line is the center of the cluster, and the broken line is the line after the center point is fitted. By observing Fig. 5a, b, it is found that as long as reasonable clustering parameters are selected, the effect of line fitting is still ideal.

Figure 6a, b compares the fitting characteristics of DB-SCAN (blue polygonal line, namely cluster) and VQ (red polygonal line, namely grid) methods. The set of points is randomly distributed according to setting III. The vertex of the broken line is the center of the cluster, and the broken line is the line after the center point is fitted. We find that both schemes can achieve the purpose of reducing path, and the clustering scheme based on grid can fully consider the discrete point set. The method of density clustering fitting path can make the path more concise.

7 Conclusion

With the development of Internet technology, crowdsourcing is used more and more widely. In order to remove the burden caused by the redundant position point set, two schemes are used to extract and fit the position data, and the advantages and disadvantages of the two schemes (DB-SCAN and Grid-K-means) are analyzed. The experimental results show that the two methods can fit the trajectory trend to some extent.

References

1. Amazon Mechanical Turk, https://www.mturk.com.
2. Google Help, https://helpouts.google.com/home.
3. Yahoo! Answers, http://answers.yahoo.com.
4. Wikipedia, http://en.wikipedia.org/.
5. Mamat T, Xie J. Research on Clustering of Agricultural Machinery Operation Trajectory Based on DBSCAN Algorithm[J]. Journal of Agricultural Mechanization Research, 2017.
6. Izakian Z, Mesgari M S, Abraham A. Automated clustering of trajectory data using a particle swarm optimization[J]. Computers Environment & Urban Systems, 2016, 55:55–65.
7. Giusto A, Piger J. Identifying business cycle turning points in real time with vector quantization[J]. International Journal of Forecasting, 2017, 33(1):174–184.
8. Calaque D, Pantev T, Toën B, et al. Shifted Poisson structures and deformation quantization [J]. Mathematics, 2017, 10(2):483–584.
9. Agustsson E, Mentzer F, Tschannen M, et al. Soft-to-Hard Vector Quantization for End-to-End Learned Compression of Images and Neural Networks[J]. 2017.
10. Valin J M, Terriberry T B. Perceptual vector quantization for video coding[J]. Proceedings of SPIE - The International Society for Optical Engineering, 2017, 9410:941009-941009-11.
11. Głomb P, Romaszewski M, Sochan A, et al. Unsupervised Parameter Selection for Gesture Recognition with Vector Quantization and Hidden Markov Models.[J]. Proceedings of the American Mathematical Society, 2017, 72(2):422–424.
12. Hasson S T, Al-Kadhum H A. Developed clustering approaches to enhance the data transmissions in WSNs[C]// International Conference on Current Research in Computer Science and Information Technology. IEEE, 2017.
13. Kang S H, Sandberg B, Yip A M. A regularized k-means and multiphase scale segmentation [J]. Inverse Problems & Imaging, 2017, 5(2):407–429.
14. Ahmadian S, Norouzi-Fard A, Svensson O, et al. Better Guarantees for k-Means and Euclidean k-Median by Primal-Dual Algorithms[C]// Foundations of Computer Science. IEEE, 2017:61–72.
15. Liu H, Wu J, Liu T, et al. Spectral Ensemble Clustering via Weighted K-Means: Theoretical and Practical Evidence[J]. IEEE Transactions on Knowledge & Data Engineering, 2017, 29 (5):1129–1143.
16. Hu C, Zhan Z, Dong K, et al. An Improved K-Means Based Design Domain Recognition Method for Automotive Structural Optimization[C]// WCX World Congress Experience. 2018.

Target Aspect Estimation in SAR Images via Two-Dimensional Entropic Thresholding and Radon Transform

Yin Long[1], Xue Jiang[1(✉)], Xingzhao Liu[1], and Zhixin Zhou[2]

[1] School of Electronic Information and Electrical Engineering, Shanghai Jiao Tong University, Shanghai 200240, China
xuejiang@sjtu.edu.cn
[2] Space Engineering University, Beijing 101400, China

Abstract. The target aspect estimation (TSE) is important to automatic target recognition in synthetic aperture radar (SAR) images, which can dramatically reduce the computational burden. In this paper, the proposed method for TSE first segments the image using two-dimensional (2D) maximum entropy threshold algorithm into target, shadow, and background. After obtaining the primary contour of target, we optimize the TSE by combining the estimation results from both Radon transform and axis information. The validity and accuracy of the proposed method are demonstrated by the experimental results with MSTAR database.

Keywords: Synthetic aperture radar (SAR) · Entropy · Aspect estimation · Target recognition · Radon transform

CLC Number: TP751

1 Introduction

Synthetic aperture radar (SAR), with its all-time and all-weather remote imaging capability, has been widely applied in earth observation and military reconnaissance. Automatic target recognition(ATR) based on SAR images has also become a hot research topic [1, 2]. However, SAR image features pose a huge challenge for automatic target recognition. The coherent imaging mechanism makes the SAR image have strong speckle noise, and the multi-view processing technique blurs the image. Segmenting the target and its shadow area from the image is the basis for the recognition work.

SAR imaging results are very sensitive to changes in imaging azimuth. Naturally, SAR image templates with different azimuth condition for each target are established, and the number of templates is very large. In practical SAR ATR processing, the approximate azimuth or azimuth range of the target to be recognized is estimated in advance. Then the template image (or extracted recognition feature) of the specific azimuth or azimuth range in the database to be identified is matched to improve the recognition efficiency and reduce the consuming time of SAR ATR [3].

© Springer Nature Singapore Pte Ltd. 2019
L. Wang et al. (eds.), *Proceedings of the 5th China High Resolution Earth Observation Conference (CHREOC 2018)*, Lecture Notes in Electrical Engineering 552,
https://doi.org/10.1007/978-981-13-6553-9_5

Many research works on target azimuth estimation of SAR images have been carried out. Principe et al. proposed an estimation based on "maximum mutual information" method [4], which is attributed to maximum likelihood method. Meth [5] achieved SAR image azimuth estimation efficiently by using the "near distance boundary". Voicu et al. [6] used outer rectangle to represent the target azimuth.

Radon transform has been applied to target azimuth estimation [7]. The basic idea is to perform Radon transform on the image contour of image to detect the straight edge on the contour, and take the maximum value of Radon transform corresponding to the direction of the straight edge (or the average value of the first few values) as the estimation of the azimuth angle of the target. Generally, the long straight edges of vehicles and other targets are consistent with the target azimuth angles, and a fine estimation can be obtained through Radon transform detection. However, the special imaging mechanism of SAR determines that the target contour is not as obvious as the optical image. In addition, under certain azimuth conditions, Radon transform may detect many straight edges on the contour due to the symmetry of the target itself and the effect of top shielding. Therefore, accurate aspect estimation cannot be obtained only based on Radon transform.

To address the above problems, we propose a new SAR aspect estimation method. First, image segmentation based on 2D maximum entropy thresholding is used to extract the accurate image contour. Then, the straight edge of the target contour is detected based on Radon transform and the corner point is detected to get the general direction of the target spindle. The final azimuth estimation is obtained by combining the Radon transform results with the target spindle information. Compared with the existing approaches, it is demonstrated that the proposed method has satisfactory estimation performance.

The rest of this paper is organized as follows. The image processing is modeled in Sect. 2. Section 3 details the proposed azimuth estimation. Experiments are conducted in Sect. 4 and conclusions are given in Sect. 5.

2 Image Processing

Since the result of image segmentation will be used for target recognition, the required accuracy is high. Due to the influence of speckle noise, SAR image segmentation based on the information provided by a single pixel does not meet the reality needs. In this paper, 2D maximum entropy thresholding is used to realize image segmentation. By making full use of the neighborhood information of each point, the accurate target segment results can be obtained.

2.1 2D Maximum Entropy Thresholding

Image maximum entropy threshold segmentation method applies the concept of entropy in information theory into image segmentation technology. Subjected to the gray statistical information constraints, the choice of threshold segments image into target area, the background area which maximizes the Shannon entropy. The common 1D maximum entropy thresholding is relatively simple, which ignores the local spatial information, and easily affected by noise. The 2D maximum entropy thresholding [8] utilizes the overall gray information reaction of pixels. It can express the image statistical information of neighbor pixels, and simultaneously has better anti-noise property. The two-dimensional histogram of the image is defined as

$$p_{i,j} = \frac{I_{ij}}{M \times N} \tag{1}$$

where $M \times N$ represent the size of image, I_{ij} denotes that the grayscale value of the image is i and the average value of the grayscale is j. Its grayscale range is $(0, L - 1)$.

Figure 1 shows a two-dimensional histogram of an image illustrating the main grayscale distribution of background and target, where the x-axis coordinates of Fig. 1b represent the mean value of the field and the y-axis coordinates represent the grayscale value distribution.

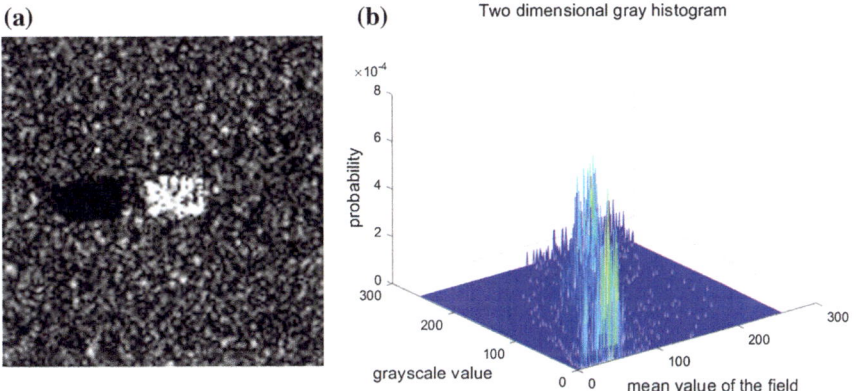

Fig. 1. The two-dimensional histogram distribution of the image. **a** original image; **b** spatial distribution of a two-dimensional histogram

Assuming O represents the ROI (region of interest) of the image, the B represents the background region, the probability P_O and P_B set at the threshold (t, s) to define regions O and B, respectively, as

$$P_O = \sum_{i=0}^{s-1} \sum_{j=0}^{t-1} p_{i,j}, \ P_B = \sum_{i=s}^{L-1} \sum_{j=t}^{L-1} p_{i,j} \tag{2}$$

Two-dimensional discrete entropy is defined as

$$H = -\sum_i \sum_j p_{i,j} \lg(p_{i,j}) \tag{3}$$

The two-dimensional entropy of region O can be obtained by normalization of probability $p_{i,j}$ of each region:

$$H_O(i,j) = -\sum_{i=0}^{s-1}\sum_{j=0}^{t-1}(p_{i,j}/P_O)\lg(p_{i,j}/P_O) = \lg(P_O) + H_O/P_O \tag{4}$$

Similarly, the two-dimensional entropy of region B can be expressed as

$$H_B(i,j) = \lg(P_B) + H_B/P_B \tag{5}$$

where H_B and H_O are given by

$$H_O = \sum_{i=0}^{s-1}\sum_{j=0}^{t-1}p_{i,j}\lg p_{i,j}, \quad H_B = \sum_{i=s}^{L-1}\sum_{j=t}^{L-1}p_{i,j}\lg p_{i,j} \tag{6}$$

Among them, the entropy of the target and the background in the whole image is defined as

$$\emptyset(s,t) = H_O(i,j) + H_B(i,j) \tag{7}$$

According to the maximum entropy principle, the optimal threshold vector (s^*, t^*) can be obtained by

$$(s^*, t^*) = \arg\max\{\emptyset(s,t)\} \tag{8}$$

2.2   Radon Transform

Given pixel $f(x, y)$, in analysis window, we compute the projection of the angle between projections direction and x-axis. The resulting projection represents the sum of the intensities of the pixels in each angle. The line of image can be expressed as

$$\rho = x\cos\theta + y\sin\theta \tag{9}$$

Radon transform function can be written mathematically by defining

$$R(\rho,\theta) = \int_{-\infty}^{+\infty}\int_{-\infty}^{+h} f(x,y)\delta(\rho - x\cos\theta - y\sin\theta)dxdy \tag{10}$$

where $\delta(\cdot)$ is the Dirac delta function. For an image, we apply image dimensions into the integral boundaries. Radon transform can be expressed as

$$R(\rho,\theta) = \int_0^Y\int_0^x f(x,y)\delta(\rho - x\cos\theta - y\sin\theta)dxgy \tag{11}$$

3   Aspect Estimation

This subsection shows details of the proposed azimuth estimation. The method contains the following steps:

(1)   Obtaining the accurate region of target based on image segmentation based on 2D maximum entropy threshold segmentation, and complete target contour can be extracted through contour curve tracking. Figure 2 illustrate an example of BMP-70 target segmentation and contour extraction results.

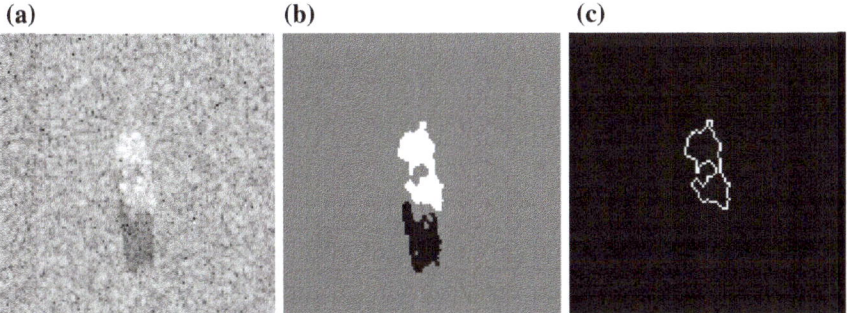

Fig. 2. Example of image segmentation and contour extraction **a** SAR image, **b** segmented target pixels and shadow pixels, **c** target contour extraction

(2)   The Radon transform was used to detect N candidate line segments with the longest length on the target contour. A Radon transform detection result is shown in Fig. 3, where $N = 3$.

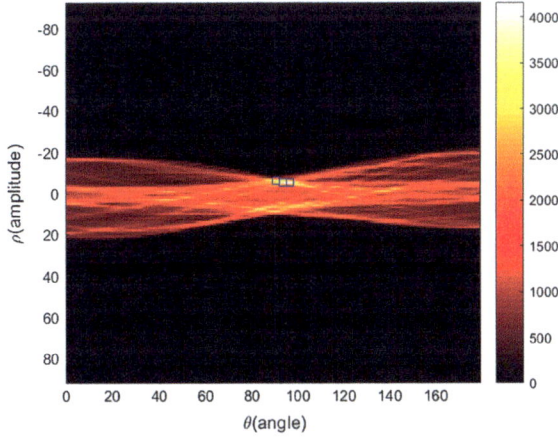

Fig. 3.   Result by Radon transform

Defining the peak value of the detection in candidate angle set η is θ_p, the maximum and minimum values of the candidate angle are θ_{max} and θ_{min}, respectively. When $\theta_{max} - \theta_{min} \leq \sigma(\sigma$ is set to 8 in this paper), the azimuth to be estimated can be express as

$$\hat{\theta} = \theta_p$$

(3) If there is a large difference between N estimates, it indicates that there are several straight edges in the target contour and the target axis information needs to be combined. In Fig. 4, the Harris detector [9] was used to obtain the corner point. Taking the line between the most distant point and the centroid of the target as the reference axis, and its directional angle is θ_A. The final aspect estimation can be obtained by

(a)   (b)

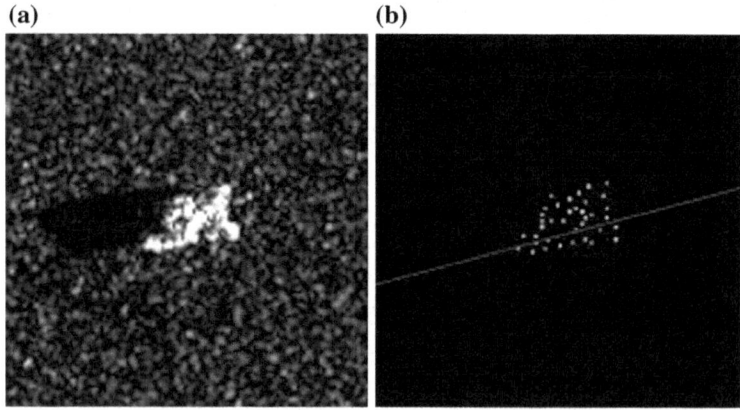

Fig. 4. Axis extraction **a** original image, **b** Harris detection and axis estimation

$$\hat{\theta} = \arg \min(|\theta - \theta_A|), \theta \in \eta$$

The processing flow based on Radon transform estimation method can be summarized as Fig. 5.

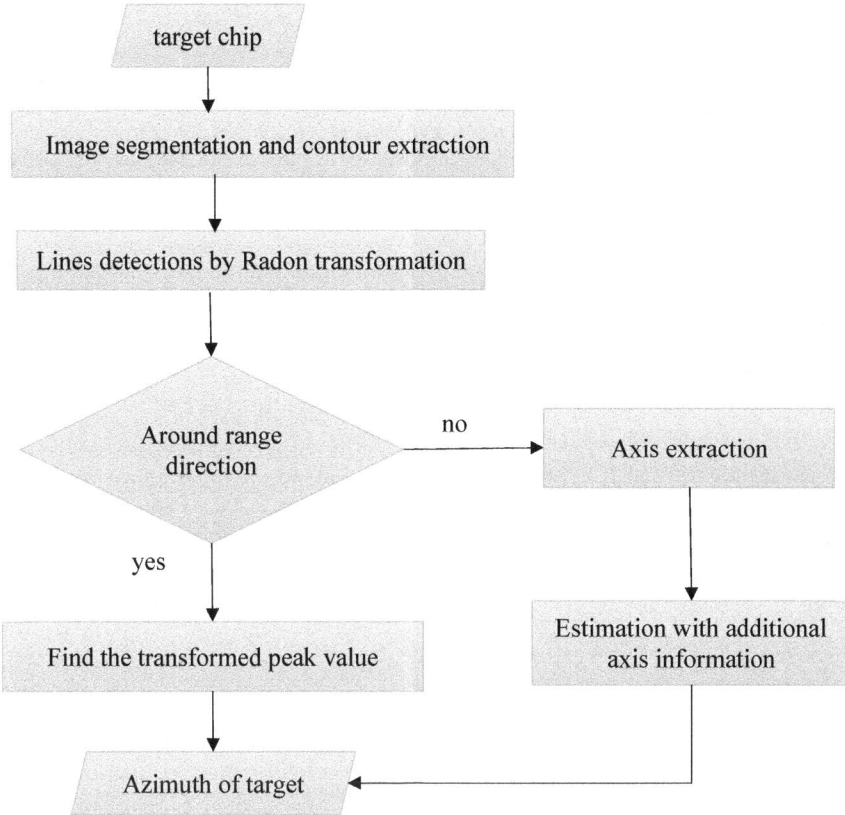

Fig. 5. Flow chart of azimuth estimation

4   Experiments

In this section, we conduct the experiments on three kinds of representative target (BMP2, BTR70, T72) in MSTAR public database (2987 targets in total). Since all azimuth estimation algorithms still have 180° of azimuth estimation ambiguity problem, the actual target aspects were wrapped to [0°, 179°]. Experimental results are illustrated in Table 1.

Table 1. Percentage of estimation with absolute error

Method	Absolute error									
	1° (%)	2° (%)	3° (%)	4° (%)	5° (%)	6° (%)	7° (%)	8° (%)	9° (%)	10° (%)
MOI	5	10	14	20	24	29	34	39	43	48
Meth	33	55	69	78	83	87	89	91	92	93
Radon transform	41	56	65	70	76	80	82	84	86	87
Proposed method	74	85	89	91	93	94	95	95	95	96

It is obvious to find the absolute error of estimations of proposed method less than other different methods which are listed in Table 1. The method obtains the best accurate estimation (such as the estimation error in the range of 1°), and correctly estimates the azimuth angle for about 96% of the target in the experiments. Besides, this method has a simpler processing flow and does not adopt the assumption that the target is a regular rectangle, which may have better performance in the azimuth estimation of asymmetric targets than method based on envelope rectangle criterion.

5   Conclusion

In this paper, a method to estimate target aspect is proposed. The 2D maximum entropy thresholding is used to segment the target and the shadow area. Meanwhile, the contour of target is extracted and aspect of the target is estimated from it by utilizing Radon transform. The azimuth estimation method is independent of the segmentation algorithm, which shows robustness. Experiment results demonstrate that the proposed method incorporates the optimal performance and simple processing flow, which will satisfy the requirements in terms of accuracy and efficiency for subsequent ATR system.

References

1. Ross T D, Bradley J J, Hudson L J, et al. SAR ATR: so what's the problem? An MSTAR perspective[C]// Aerosense. International Society for Optics and Photonics, 1999.
2. Jones, Grinnell, and B. Bhanu. "Recognition of Articulated and Occluded Objects." *Pattern Analysis & Machine Intelligence IEEE Transactions on* 21.7(1999):603–613.
3. Principe, Jose C, Q. Zhao, and D. Xu. "A novel ATR classifier exploiting pose information." *Proceedings of Image Understanding Workshop* (1998).
4. Principe, Jose C., D. Xu, and J. W. F. Iii. "Pose estimation in SAR using an information theoretic criterion." *Proc. SPIE Conf* 1998.
5. Meth, Reuven. "Target/shadow segmentation and aspect estimation in synthetic aperture radar imagery." *Proceedings of SPIE—The International Society for Optical Engineering* (1998):188–196.

6. Voicu, Liviu I, R. Patton, and H. R. Myler. "Multicriterion vehicle pose estimation for SAR ATR." *Proc Spie* 3721(1999).3321. Bellingham, WA: society of Photo-Optical Instrumentation Engineers, 1999: 497–506.
7. Zhang, Xuepan, *et al.* "Geometry-Information-Aided Efficient Motion Parameter Estimation for Moving-Target Imaging and Location." *IEEE Geoscience & Remote Sensing Letters* 12.1 (2017):155–159.
8. Chen, Meng Xian, *et al.* "Improved Artificial Colony Algorithm in the Application of Two-Dimensional Otsu Image Segmentation." *Computer Systems & Applications* (2014).
9. Ryu, J. B, C. G. Lee, and H. H. Park. "Formula for Harris corner detector." *Electronics Letters* 47.3(2011):180–181.

Signal Model and System Parameter Study for Geosynchronous Circular SAR

Bingji Zhao[1(✉)], Qingun Zhang[1], Chao Dai[1], Han Li[2], Zheng Lv[1], and Jian Liang[1]

[1] Beijing Institute of Spacecraft System Engineering CAST,
Beijing 100094, China
zachary_zbj@163.com
[2] National Laboratory of Radar Signal Processing, Xidian University,
Xi'an 710071, China

Abstract. This manuscript illustrates some key problems about Geosynchronous circular SAR system. Geo-CSAR possesses the potential ability of 3D surface resolving power through variant angle of view. Due to especial requirement of the satellite trajectory, the Geo-CSAR satellite's six orbital key parameters could be calculated and designed detailedly. Furthermore, this paper illuminates a 3-D imaging signal model of Geo-CSAR during its long integration time according to the satellite's curve trajectory. Lastly, a designing approach of its key system parameters is proposed in this manuscript. A simulation experiment is performed to verify the efficiency and superiority of this approach, and the results represent that it has a good effect on an L-band mode Geo-CSAR tomography system with elevation resolution higher than 5 m.

Keywords: Geo-SAR · Circular SAR · Signal model · System parameters

1 Introduction

The feasibility of global mapping by low earth orbital (Leo-SAR) has been validated, for instance GF-3, by the current spaceborne SAR systems. Geosynchronous Earth orbital Synthetic Aperture Radar (Geo-SAR), firstly proposed by Tomiyasu in 1978, is characterized by short revisit period and large observing swath. It shows potential for many different applications such as earthquake monitoring, geologic imaging, natural disasters prevention, and so on.

As its orbital altitude is about 35,790 km, Geo-SAR is different from Leo-SAR, and its period is equal to one Earth day (86,164 s). Due to the very principle of SAR that the relative motion between the radar and the targets under observation is necessary, it requires inclined orbits. The nadir trace can be presented as figure "8", circle, line, dripping form, and so on, when its orbital inclination angle, eccentricity and perigee argument accord with some regulation [1–3].

Circular SAR(CSAR), advanced first by D. G. Falconer and G. J. Moussally in 1995, is developed as an innovative SAR system in recent years. Obtaining the dispersion

© Springer Nature Singapore Pte Ltd. 2019
L. Wang et al. (eds.), *Proceedings of the 5th China High Resolution Earth Observation Conference (CHREOC 2018)*, Lecture Notes in Electrical Engineering 552,
https://doi.org/10.1007/978-981-13-6553-9_6

information from several different azimuthal angles along the flight track, CSAR can obtain more target specifics and achieve higher three-dimensional (3D) resolution. Compared with conventional SAR, CSAR can improve Signal-to-Noise Ratio (SNR) and reduce altitude ambiguity, which makes it could be employed by lots of applications, such as military reconnaissance and terrain monitoring. It is impossible for the conventional Leo-SAR to form a circular aperture, however, a circle satellite trajectory can be traced out through designing appropriate geosynchronous orbit parameters, including orbit inclination, perigee argument, and eccentricity. As a result, the circular SAR could be connected with Geo-SAR in the circular nadir trace condition. Geo-CSAR system has many advantages which make it superexcellent remote sensing instrument in the future. It can provide unique large coverage high-resolution three-dimensional 3D images. Besides, it has continuous 24-h observing capability [4, 5].

This manuscript discusses some key problem of Geo-CSAR. First, an orbital parameter designing approach to form circular nadir trace is advanced. The geosynchronous satellite movement could be similarly considered as two period anharmonic vibrations in longitude and latitude direction separately. Circular orbital parameter calculation expressions are listed based on it. Second, the Geo-CSAR geometry model and signal model are illustrated. And a 3-D resolution property is analysed based on these models. In ideal condition, the CSAR imaging resolution is made with full 360° circular apertures measurements. As the altitude resolution is decided by synthetic angle, the signal model accuracy is discussed in the long integration time. Thirdly, some key parameters of a L-band mode Geo-CSAR tomography system with 3-D resolution higher than 5 m are designed in this paper. Finally, some simulation results are shown to validate the conclusions in this paper.

2 Geometry Model and Orbital Designing of Geo-CSAR

2.1 Geo-CSAR Geometry Model

Geo-CSAR could be considered as a special spotlight mode SAR. It obtains the 3-D terrain information provided by spotlight mode circular synthetic aperture radar. The circular synthetic aperture refers to a single circular flight path of 360° azimuth at a constant altitude and broadside squint angle in ideal condition. Be similar to the conventional synthetic aperture in flighting direction, high two-dimensional resolution could be obtained by synthetic aperture. A synthetic aperture could be similarly formed in different azimuthal angles directions along the circular flight track, and the tomography 3-D resolution in elevation direction can be got. A Geo-CSAR tomography system geometry model is drawn as follows.

As depicted in Fig. 1, the geometry model is advanced in the Earth Rotation Coordinate (ERC) O-XYZ, in which symbol O, the earth centroid, represents the coordinate origin, X-axis, witch circumvolves with earth rotation, points to local Greenwich longitude direction, the Z-axis is along the earth north pole direction, and the Y-axis obeys the Cartesian right-hand rule. S_o is the point of intersection of Z axis and satellite fight plane, E_o is the point of intersection of Z axis and Earth surface, θ is the look angle, R_e is the Earth radius, and H is the satellite height.

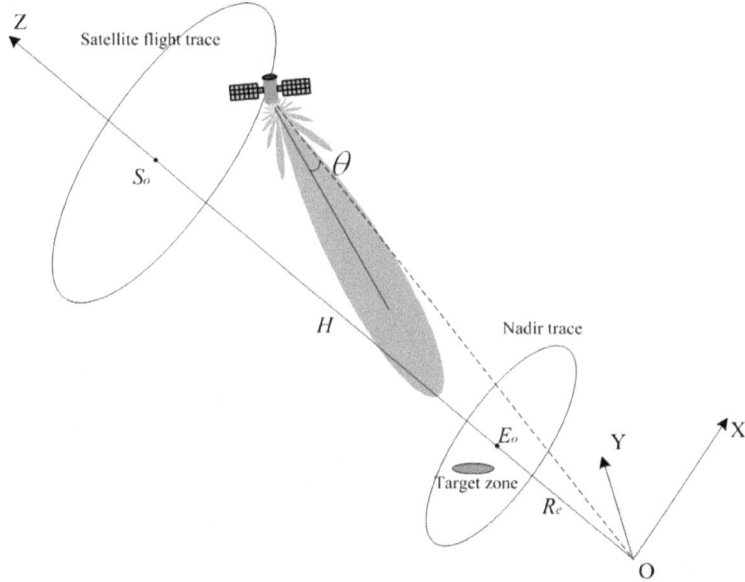

Fig. 1. Geo-CSAR tomography imaging system geometry model

2.2 Orbital Parameter Designing of Geo-CSAR

Generally, an ideal model, defined as Kepler Orbit Elements, is advanced to illustrate the geometry problems of spaceborne SAR. The satellite's orbit is basically decided by six orbital parameters, and WGS-84 earth model is employed for analysing in this paper. Thus, a two-body model is presented to describe the satellite movement property with the mass gravitational of homogeneous earth. The perturbation aspects (mainly include the harmonic gravity due to Earth's non-spherical earth perturbation, atmospheric resistance, the moon and the sun's gravity, etc.) are all ignored in this model. The classic satellite orbital equation is expressed as follows.

$$R_\mathrm{s} = \frac{a(1 - e^2)}{1 + e\cos f} \tag{1}$$

where R_s is the range between satellite centroid and earth core, a is the half long axis, e is the eccentricity, f is the true anomaly, and it could be calculated as follows:

$$\tan\frac{f}{2} = \sqrt{\frac{1 + e}{1 - e}}\tan\frac{E}{2} \tag{2}$$

$$M = n(t - \tau_\mathrm{p}) = E - e\sin E \tag{3}$$

$$E = M + e\left(1 - \frac{1}{8}e^2 + \frac{1}{192}e^4\right)\sin M + e^2\left(\frac{1}{2} - \frac{1}{6}e^2\right)\sin 2M + e^3\left(\frac{3}{8} - \frac{27}{128}e^2\right)\sin 3M$$
$$+ \frac{1}{3}e^4\sin 4M + \frac{125}{384}e^5\sin 5M$$

$$(4)$$

where E is the eccentric anomaly, M is the mean anomaly, n is mean angle velocity, τ_p is the time when satellite pass through perigee, and t represents the orbital time. The satellite nadir longitude and latitude could be developed according to the comment above.

$$\begin{cases} \text{Lat} = \arcsin[\sin(f + \omega)\sin\theta_i] \\ \text{Long} = \arctan[\tan(f + \omega)\cos\theta_i] + \Omega - \omega_e t \end{cases} \quad (5)$$

where Lat and Long separately mean the nadir's latitude and longitude, w is the perigee argument, Ω is the ascending longitude, w_e is the mean angle velocity of earth rotation, and θ_i is the orbital inclination. Different geosynchronous satellite' nadir traces could be drawn separately as Fig. 2. We can see from Fig. 2 that the nadir trace patterns are

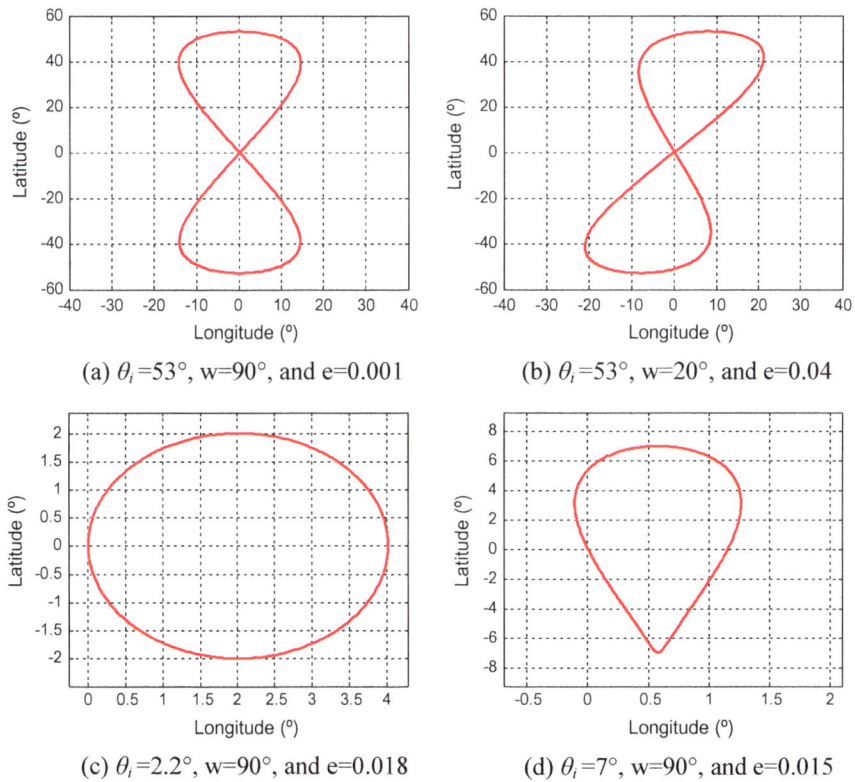

(a) $\theta_i=53°$, w=90°, and e=0.001

(b) $\theta_i=53°$, w=20°, and e=0.04

(c) $\theta_i=2.2°$, w=90°, and e=0.018

(d) $\theta_i=7°$, w=90°, and e=0.015

Fig. 2. Nadir trace by different orbital parameters

figured as "8" type, slant "8" type, circle type, and drip type, while the perturbation is ignored. In fact, its nadir trajectory sort could be many and various. The circle trajectory type could be obtained when the e is far below 1 and θ_i is much less than π. As a result, the expression (6) could be similarly derived as Eq. (6) while the formulas (2)–(4) are substituted into (5).

$$
\begin{cases}
\text{Lat} \approx 2e\, \sin\big[\omega_e(\tau - \tau_p)\big] + \omega + \Omega \\
\text{Long} \approx \theta_i \sin\big[\omega_e(\tau - \tau_p) + \omega\big]
\end{cases}
\tag{6}
$$

According to the content above, it could be approximatively considered that the nadir moves along the longitude and latitude direction by simple harmonic motion. Thus, the parameter for a circle type could be calculated by expression (5).

3   Signal Model Analysis

While the transfer signal is modulated as LFM form, the single point target echo signal of Geo-CSAR could be described as follows.

$$
S_0(\tau, \eta) = K \int_V g_a(r_t, r_s)\rho(r_t) \exp\left\{ j\pi K_r \left(\tau - \frac{2R_t(\eta)}{c} \right)^2 \right\} \exp\left\{ -j\frac{4\pi f_0 R_t(\eta)}{c} \right\} R_t^{-2} dr_t
\tag{7}
$$

where τ is fast time, η is slow time, R_t is the range between the target and the radar position, r_t and r_s separately represent the position vector of target and satellite, $g_a(r_t, r_s)$ is the antenna gain function, and the factor K means the remaining parameters of the radar equation. While a matched filter is employed, a conjugate reference function multiply is processed, and the Eq. (7) can be presented as:

$$
S_0(\tau, \eta) = K \int_{\theta_s}^{\theta_e} g_a(r_t, r_s)\rho(r_t) \cdot \exp\left\{ -j\pi K_r \left(\tau - \frac{2R_0(\eta)}{c} \right)^2 \right\}
$$
$$
\exp\left\{ j\pi K_r \left(\tau - \frac{2R_t(\eta)}{c} \right)^2 \right\} \exp\left\{ -j\frac{4\pi f_0 R_t(\eta)}{c} \right\} R_t^{-2} dr_t
\tag{8}
$$

where θ_s and θ_e separately represent the imaging start and end angle in the flight circle, R_0 is the refer range between the refer target and the satellite. After series of complicated deducing, the above echo signal expression could be rewritten as:

$$
S_0(r_t, r_0) = \rho(r_t) \int_{\theta_s}^{\theta_e} \sin c\left[\frac{2B(R_t - R_0)}{c} \right] \exp\left\{ -j\frac{4\pi f_0 (R_t - R_0)}{c} \right\} d\theta
\tag{9}
$$

Assume that a full 360° circular aperture measurement is processed, the equation above could be presented as:

$$S_0(r_t, r_0) = \sin c\left[\frac{2B(R_t - R_0)}{c}\right] \int_{-\pi}^{\pi} \exp\left\{-j\frac{4\pi}{\lambda(R_s - Z_s)} R_c \rho_{ar} \cos\theta_c\right\} d\theta \qquad (10)$$

where R_s is the distance from satellite to earth core, Z_s is the z-axis coordinate of satellite, R_c is the radium of the circular aperture, ρ_{ar} could be considered as the two-dimensional resolution, and θ_c represents the coherent angle. Thus, the three-dimensional focusing result of the Geo-CSAR can be advanced as:

$$S_0(r_t, r_0) = \sin c\left[\frac{2B(R_t - R_0)}{c}\right] J_0\left(\frac{4\pi}{\lambda(R_s - Z_s)} R_c \rho_{ar}\right) \qquad (11)$$

where $J_0(*)$ represents the zero order Bessel function. The expression (11) is obtained in full 360° circular aperture condition. Furthermore, a more complicated result, which is got in partial circular aperture condition, could be described by high order Bessel function. As the long integration time, the slant range of Geo-CSAR should be described by high order Doppler range equation (DRM) as follows.

$$\begin{cases} R_{DRM}(\eta) = R_c + \sum_{n=1}^{\infty} k_n(\eta - \eta_c)^n \\ k_1 = -\frac{\lambda}{2} f_{dc} \\ k_n = -\sum_{n=2}^{\infty} \frac{\lambda}{2n!} f_{(n-1)r}, \quad n \geq 2 \end{cases} \qquad (12)$$

The expression above provides a universal approach to calculate accurate slant range needed in Eqs. (7)–(10). The order number can be decided by circular aperture angle and orbital parameter.

4 System Parameter Designing and Simulation

4.1 Geo-CSAR System Parameter Designing

The main parameter of an L-band mode Geo-CSAR tomography system with 3-D resolution higher than 5 m is discussed in this section.

(a) Orbital parameters

According to expression (5), the circle trajectory type is possible if e is far below 1 and θ_i is much less than π. However, the orbital inclination should be as big as possible to obtain large imaging swath. A 14° inclination and 0.065 eccentricity orbit is chosen. The satellite and nadir trajectory are shown as follows (Fig. 3).

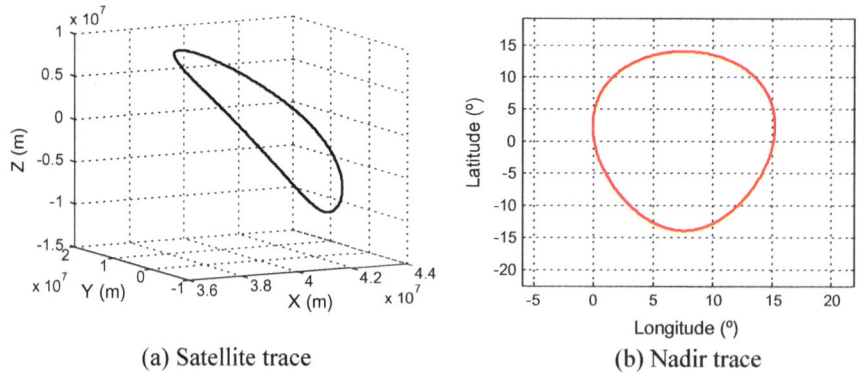

(a) Satellite trace   (b) Nadir trace

Fig. 3. A circular geosynchronous satellite orbit

(b) System bandwidth and PRF

The radium of the satellite circular aperture R_c can be calculated as 20,200 km, and the radium of nadir circular is 1560 km. The system bandwidth decides the elevation resolution, and it can be obtained as:

$$B_r = \frac{c}{2\rho_z} \tag{12}$$

To achieve the 5 m elevation resolution, the bandwidth should be wider than 30 MHz. System PRF should be designed to receive the whole target area echo data and to avoid the elevation ambiguity. A 5 m resolution needs at least 60 Hz Doppler bandwidth, and the nadir circular scale echo data needs longer than 0.008 ms PRT. Thus 110 Hz PRF is selected.

(c) Antenna aperture and NEXZ

The Geo-CSAR system NEXZ can be calculated as follows:

$$NE\sigma^0 = \frac{\sigma^0}{SNR} = \frac{(4\pi)^3 r^4 L_s (kT_0 B_n F_n)}{P_t G^2 \lambda^2 \rho_{gr}\rho_a T_{sar} PRF \cdot T_r B_r} \tag{13}$$

A 20 m radium round antenna is selected, and its radiation peak power is assumed as 20,000 W. The beam centre look angle is assumed as 2.6°, and the NEXZ can be simulated as follow. The beam centre value is about −31.5 dB (Fig. 4).

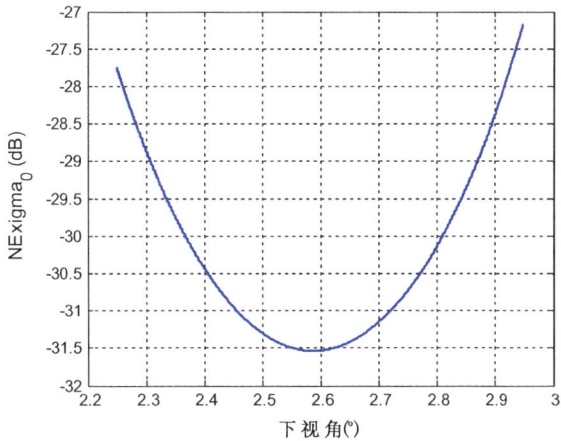

Fig. 4. The NEXZ simulation result of Geo-CSAR

(d) Main parameters

Section (a), (b), and (c) analyses some main parameters in details. Anyway, Geo-CSAR's core parameters are listed as follows (Table 1).

Table 1. Key parameters of Geo-CSAR

Name	Value	Name	Value
Half long axis	42164 km	Peak power	20,000 W
Inclination	14°	Pulse width	200 μs
Eccentricity	0.065	Occupy rate	15%
Perigee argument	90°	Antenna size	20 m
Earth model	WGS-84	NEXZ	−31.5 dB
Bandwidth	50 MHz	Elevation resolution	≤5 m
PRF	110 Hz	Frequency	1.50 GHz

4.2   Point Target Imaging Simulation

In this section, a point target is employed to validate the theory above. The imaging result is shown as follows. As depicted in Fig. 5, a zero-order Bessel function pattern focusing result is obtained. The elevation resolution is 3.16 m, and the azimuth and range resolution is 0.04 m.

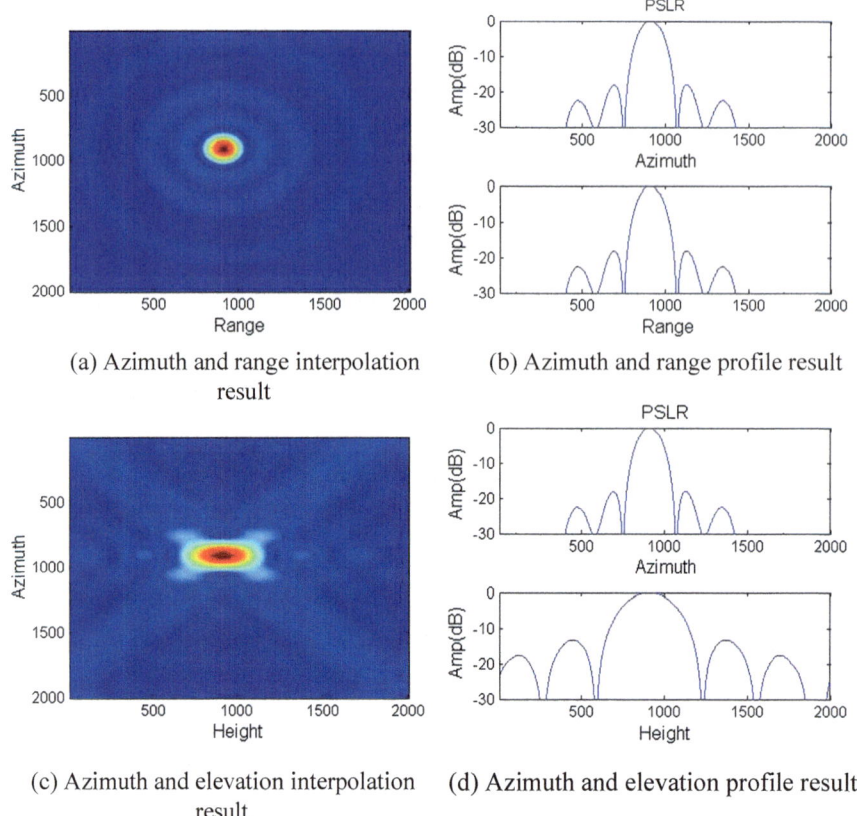

(a) Azimuth and range interpolation
result

(b) Azimuth and range profile result

(c) Azimuth and elevation interpolation
result

(d) Azimuth and elevation profile result

Fig. 5. Point target imaging results

5   Conclusion

Some key problems about geosynchronous circular SAR are discussed in this letter. Due to its particular requirement of the satellite trajectory, which is a circular aperture, the analytic calculation approach is given in Sect. 2.1, and its six orbital key parameters that answer for needs can be obtained. Then the 3-D imaging signal model during long integration time is illuminated in Sect. 3. In Sect. 4, the designing approach of its key system parameters is proposed. And a simulation experiment is performed to verify the efficiency and superiority of point of view in this paper, and the results show that it has a good effect on an L-band frequency Geo-CSAR system with 3-D resolution around 3.16 m by point target imaging simulation experiment.

Acknowledgements. Thanks for the support from the national natural science foundation of China 'Analysis on the non-ideal factors in Geo-SAR tomography imaging' (ID: 61601022) and the national key R&D Program (ID: 2017YFB0502700).

References

1. Ian G Cumming and Frank H Wong, "Digital Processing of Synthetic Aperture Radar Data: Algorithms and Implementation"[M]. Boston, Artech House, 2005.
2. K. Tomiyasu, J. L. Pacell, "Synthetic Aperture Radar Imaging from an Inclined Geosynchronous Orbit", IEEE Transaction on Geoscience and Remote Sensing, 1983, 21, (3), pp. 324–328.
3. Xinwu Li, Lei Liang, Huadong Guo, and Yue Huang, "Compressive Sensing for Multibaseline Polarimetric SAR Tomography of Forested Areas"[J], IEEE Transaction on Geoscience and Remote Sensing, vol. 54, no. 1, January, 2016.
4. Kou Leilei, Wang Xiaoqing, Xiang Maosheng, Chong Jinsong, et.al. "Circular SAR processing using an improved omega-k type algorithm"[J]. Journal of Systems Engineering and Electronics. 2010, 21(4):572–579.
5. Xiao Wang, Feng Xu, and Ya-Qiu Jin, "The Iterative Reweighted Alternating Direction Method of Multipliers for Separating Structural Layovers in SAR Tomography" [J], IEEE Geoscience and Remote Sensing letters, vol.14, no. 44, November, 2017.

Low-Light Remote Sensing Images Enhancement Algorithm Based on Fully Convolutional Neural Network

Wuzhen Jian[1,2], Hui Zhao[1(✉)], Zhe Bai[1], and Xuewu Fan[1]

[1] Xi'an Institute of Optics and Precision Mechanics, Chinese Academy
of Science, Xi'an, Shaanxi 710119, China
`optic.jian@foxmail.com`, `zhaohui@opt.ac.cn`
[2] Shaanxi Normal University, Xi'an, Shaanxi 710100, China

Abstract. Low-light remote sensing is a powerful complement to daytime optical remote sensing and can greatly expand the time domain of high-resolution earth observations, and make day and night imaging possible. However, when a low-light sensor is used in the morning dusk and dawn, the captured images have characteristics of low contrast, low brightness, and low signal-to-noise ratio, which severely restrict the identification and interpretation of ground objects. Traditional low-light image enhancement algorithms such as histogram equalization, gamma conversion, and contrast-limited adaptive histogram equalization algorithm, and so on can enhance the low-light remote sensing image and solve the problem of contrast enhancement, but the noise amplification effect brought by the enhancement will degrade the signal-to-noise ratio of the enhanced image. Therefore, in this paper, a data-driven low-light remote sensing image enhancement algorithm is studied. First of all, lots of low-light raw image data pairs corresponding to very low illumination are captured. Then, these raw image data are used to train a deep fully convolutional neural network composed of an encoder–decoder structure. After that, the low-light remote sensing images could be enhanced by the pretrained net structure. The numerical results demonstrate that the fully convolutional neural network based on enhancement algorithm greatly improves the brightness and the contrast of low-light images compared with the traditional enhancement algorithms while a high enough signal-to-noise ratio could be preserved, which will make interpretation and identification much easier.

Keywords: Low-light remote sensing image enhancement · Fully convolutional neural network · HE · Gamma correction · CLAHE · Median filtering

1 Introduction

Low-light remote sensing imaging is an advanced imaging technology based on low-light sensors (such as EMCCD, ICCD, SCMOS, etc.) to achieve remote sensing imaging [1, 2] at low-light condition. It plays an increasingly important role in research and application field of ecological environment research in low-light environment,

L. Wang et al. (eds.), *Proceedings of the 5th China High Resolution Earth Observation Conference (CHREOC 2018)*, Lecture Notes in Electrical Engineering 552,
https://doi.org/10.1007/978-981-13-6553-9_7

natural disaster monitoring and disaster emergency warning, urban construction and informatization, global sea surface monitoring, extreme weather monitoring and early warning, nighttime meteorological observation, and so on. Although the low-light detector solves the problem of limited sensitivity of the detector, the raw low-light remote sensing image often demonstrates low brightness, low contrast, low signal-to-noise ratio, narrow dynamic range, and blurred visual effects [3–5], which is bad for identification and interpretation. Therefore, how to improve the quality of low-light remote sensing images and improve the corresponding visual quality are the issues that will be addressed in this paper. It will greatly enhance the data application performance of low-light remote sensing images.

The traditional enhancement algorithms for low-light remote sensing images can be roughly divided into two categories. One is carried out in spatial domain and the other is implemented in transform domain. Nearly, all the spatial domain enhancement algorithms including histogram equalization (HE) [6], adaptive histogram equalization (AHE) [7], contrast-limited adaptive histogram equalization (CLAHE) [8] and gamma conversion (gamma correction) [9], etc., are based on direct manipulation of pixels. In HE, as this classical algorithm applies the same transform to the entire image, it cannot adapt to the contrast variation of different regions. Especially, when the image includes a very dark region, the brightness will be excessively increased but the grayscale dynamic range of the entire image has not been effectively improved. Besides that, partial combination of effective pixels caused by HE can also lead to loss of image information. The improved AHE and CLAHE algorithms are designed based on the classical histogram equalization algorithm. Both AHE and CHAHE calculate the histogram of multiple local regions of the image and redistribute the brightness to change the image contrast. However, AHE tends to increase the local contrast too much and makes image distortion sometimes. Although CLAHE performs better, but it will amplify noise for very low illumination images after enhancement. The gamma transform extends the brightness of dark regions in low gray levels by compressing pixels of high gray levels, but it cannot avoid the noise amplification effect either.

The image enhancement algorithms implemented in the transform domain mainly includes wavelet transform [10, 11], Retinex algorithm [12, 13], dark channel prior algorithm [14, 15], and so on. Most of these algorithms have three characteristics in common. First, they assume that the images already contain good representations of the scene content. Second, they do not explicitly simulate the noise of the image. Third, they usually require further noise reduction processing, for example, median filtering [16, 17], Gaussian low-pass filtering [18], NLM [19], BM3D [20, 21], etc. Since the imaging under extremely low illumination have severe noise, the above-conventional method can improve the brightness and contrast of the low-light image to a certain extent, but the effect of suppressing noise amplification is not satisfactory.

Therefore, in this paper, a neural network-based low-light image enhancement algorithm is proposed and studied. Considering that low-light image enhancement and denoising can be seen as a problem of mapping a noisy low-light image to a noise-free image with normal illumination (reference image), the image enhancement and denoising process can be regarded as the fitting process of the mapping, and the neural network has strong mapping approximation ability. Compared with the traditional low-light image enhancement and denoising algorithms, this algorithm has three advantages. First, it is a highly complex nonlinear system that can realize nonlinear mapping between input data and output data. Second, it can solve problems by learning and can adapt to environmental changes. Third, it can be implemented in parallel which is suitable for fast real-time processing.

Studies have shown that the deep fully convolutional neural network can be used to fit the low-light image enhanced denoising process, and achieve enhancement and denoising effects [22, 23] at the same time. Therefore, this paper proposes a data-driven approach to solve the enhancement-denoising problem of very low illumination images. This approach is based on the bio-image segmentation U-net [24] network architecture and is to be expanded in this paper. This end-to-end training process can effectively avoid noise amplification and error accumulation, which is far better than traditional methods. It will significantly improve the visual quality of the low-light remote sensing images and will make interpretation and identification much easier.

2 Method

2.1 Description of Fully Convolutional Neural Network Structure

Figure 1 shows a detailed structure about the fully-convolutional neural network. Each gray cylinder represents a multichannel feature map, the number above it represents the number of channels of the feature map, and the white cylinder represents the feature map copied from the left. The number on the leftmost edge of each cylinder represents the dimension size of the feature map, and the arrows of different colors indicate different operations on the feature map.

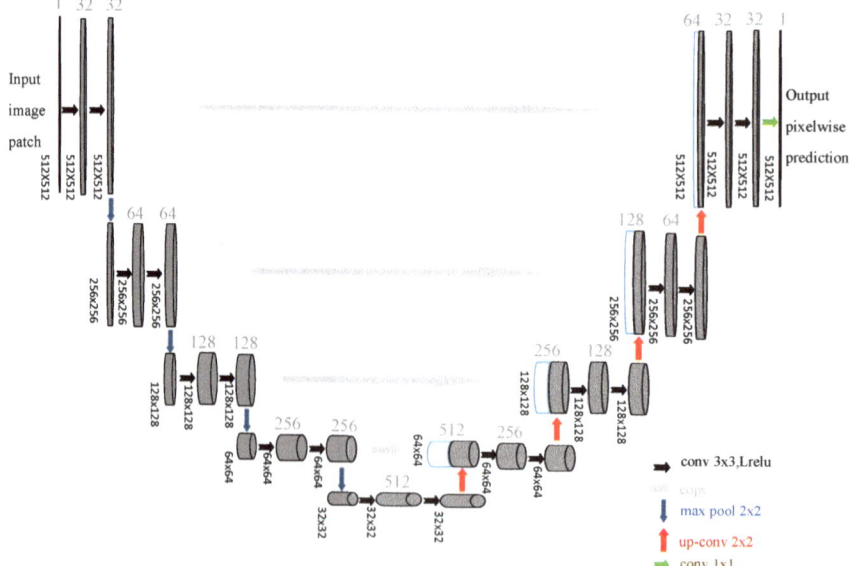

Fig. 1. Fully ConvNet structure

The network structure can be seen as a 23-layer network consisting of two parts: encoder and decoder. They separately realize feature decomposition and low-contrast image reconstruction for low-light image context information. Finally, the low-light image is reconstructed by predicting each pixel. The leftmost part is the image coding structure, which is a typical repetitive convolutional network structure with two convolutional layers and one pooling layer in each iteration. In order to make the size of the feature map unchanged, after each roll operation, the convolution kernel size of the convolutional layer is set to 3×3, the step size is 1, the number of zero padding is 2, and the Lrelu activation function is used after finishing convolution operation each time. Followed by a 2×2 size maxpool layer with a step size of 2, the feature map size is reduced to 1/4 of the original image after each downsampling operation, but the number of feature map channels is doubled. On the one hand, this procedure is to reduce the amount of calculation. On the other hand, it is better to extract deeper features.

The rightmost part is the image decoding structure. In fact, the deconvolution operation is used instead of the pooling layer operation at the back of the network. The network structure is basically symmetrical to the coding structure and the deconvolution layer will increase the resolution of the output image. In each deconvolution, the feature map size is doubled and the number of channels is reduced by half. In order to make the reconstructed image information as complete as possible, during the image coding contraction process, it will generate the high-resolution local feature map information and this information will be merged into the feature map with the result of the deconvolution operation. Then, the convolution operation is performed twice behind the merged feature map so as to extract more abstract features. After a series of deconvolution operations, the restructured feature map has the same size as the input image. Finally, the 1×1 size convolution kernel is used to reduce the dimension of the 32-channel feature map into a one channel feature map that we need.

2.2 Preparation of Dataset Used for Training

The SID dataset [25] is used for neural network training, which is obtained under very low illumination and retains the raw image data information. Since the low-light remote sensing image is panchromatic imaging, colors are not considered here. So, the single channel is used as the input data for network training. Because the environment illumination is about 0.2–5 lx and the exposure time is short, most of the images will look very dim. In the same scene, each low-light image has a corresponding long-exposure high-quality image (named as ground-truth reference image), and each long-exposure image corresponds to multiple short-exposure images.

Table 1 below lists the image dataset distribution information with different short-exposure times:

Table 1. Dataset distribution

Amplification	Exposure times (s)	#Iimages
×300	1/10,1/30	1190
×250	1/25	699
×100	1/10	808

As can be seen from Table 1, the dataset contains a total of 2697 raw short-time exposure images. From leftmost column to rightmost column, the contents listed in Table 1 are the exposure time ratio of the input image and the reference image, the exposure time of the input image, and the number of images in each case, respectively. Figure 2 displays partial image samples in the SID dataset. Ground-truth pictures are displayed at the front, and the short-time exposure input images are displayed behind.

Fig. 2. Partial sample images in the dataset

2.3   Training Process of Fully Convolutional Neural Network

As can be seen from Fig. 3, it achieves extremely low illumination image enhancement. Since the low-light image has a characteristic of directional invariance, the change in image orientation does not affect the prediction of the image. Therefore, this paper adopts data enhancement methods including horizontal flipping, vertical flipping, random cropping, rotation, and scale transformation. It can not only expand the number of training samples, but also increase the diversity of training samples. By doing this, on the one hand, it can avoid over-fitting and on the other hand, it can improve the performance of the model, and make the network more robust.

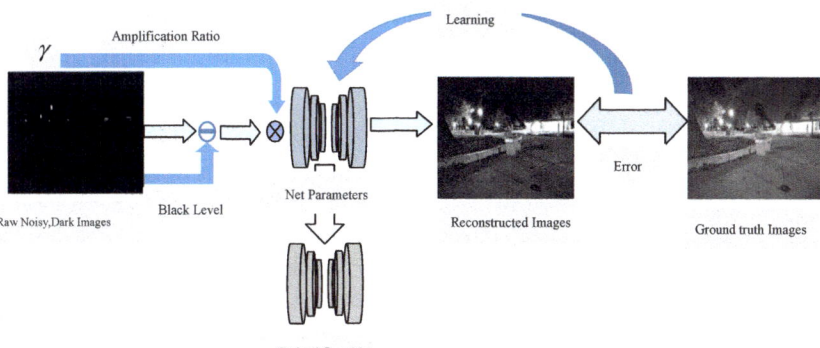

Fig. 3. The structure of low-light image processing pipeline

The raw image data are first packed into one channel. After that, the black level is subtracted from the packed raw image data. Finally, the input data is scaled by using the expected amplification ratio (For example, ×250, ×300). Then the packed and amplified input data are fed into a fully convolutional neural network. The network's weight parameters and offsets were iteratively updated continuously and the error between the reconstructed image and the reference image is getting smaller and smaller.

2.4   Neural Network Training Based on Adam Optimizer

In this paper, the L1 loss function and l_2 regularization are used as the objective function of the error calculation between the input image and the reference image. The function expression is shown in Eq. (1):

$$L(w, b) = \frac{1}{2N} \sum_{i=1}^{N} |\hat{y}_i(w, b) - y_i| + \frac{\lambda}{2} \|w\|^2 \tag{1}$$

where the N is training sample number, the w is weight parameter, and the b is offset parameter; $\hat{y}_i(w, b)$ is the predict recombined image for the number i sample, λ is the weight penalty factor.

Adam optimizer [26] is adopted as the neural network iterative optimizer. During each iteration of training, the input data of the network is a single raw short-exposure image. Then, the objective function is used to calculate the error between the output by the trained neural network and the ground-truth image. After that, Adam optimizer is applied to optimize the average error to update the network parameters. When training and testing images, the amplification ratio settings are determined based on the ratio of exposure times between short-exposure input images and corresponding long-exposure reference (ground truth) images, for example, ×300, ×200, and ×100. During each iteration, we randomly cropped the 512 × 512 image block and randomly mirrored it and flipped it into the neural network training. The learning rate was set to 0.0001 at the beginning, and then gradually decreased it after a series of iteration. When the final loss error value is reduced to 0.03, the training is terminated.

3 Analysis of Numerical Results Based on SID Dataset

The comparison of the enhancement results corresponding to the traditional enhancement algorithm and the algorithm proposed in this paper is demonstrated in Fig. 4. At the same time, the histogram of enhanced images corresponding to each kind of algorithm is given synchronously in Fig. 4. In Fig. 4, (a) is the original low-light image and its histogram after subtract the black level and multiplied the amplification ratio; (b) is the histogram equalization effect and its histogram; (c) is the gamma transformation effect and its histogram; (d) is the CLAHE enhancement algorithm's effect and its histogram; (e) is our result effect and its histogram; (f) is the ground truth and its histogram. Besides that, The PSNR [27] is used to assess the effectiveness of the proposed algorithm, as shown in Table 2.

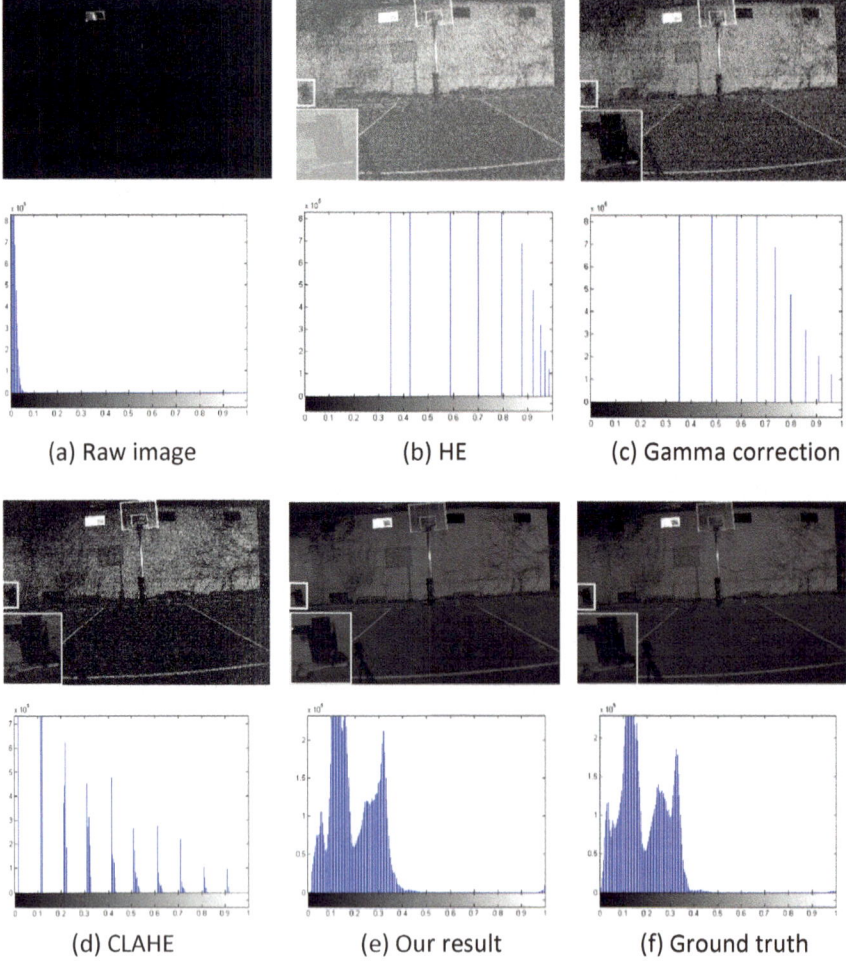

Fig. 4. Comparison of different enhancement methods for low-light image enhancement

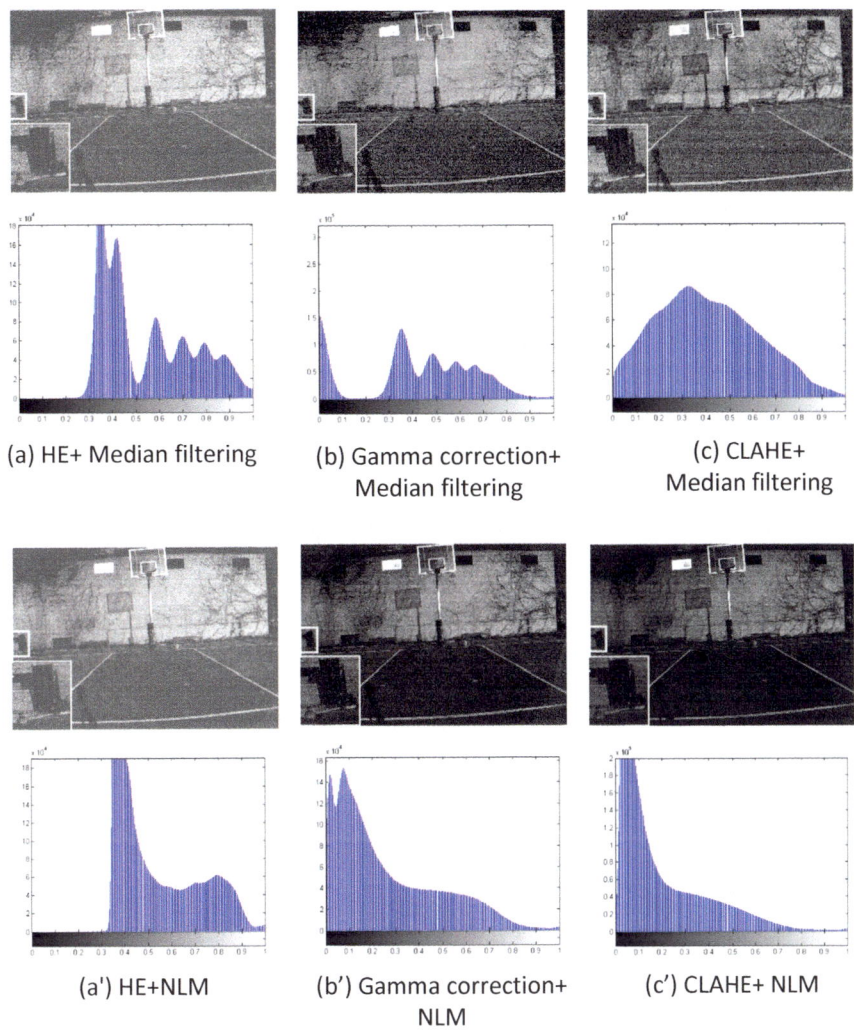

(a) HE+ Median filtering   (b) Gamma correction+   (c) CLAHE+
                              Median filtering   Median filtering

(a') HE+NLM   (b') Gamma correction+   (c') CLAHE+ NLM
                NLM

Fig. 5. The effects of different denoise method after the low-light image enhancement

Table 2. PSNR in each condition

	HE	Gamma correction	CLAHE	Our result
No denoise	5.3661	10.0097	10.5335	**27.1452**
Median filtering	8.4481	11.3314	11.7863	—
NLM denoise	13.4512	15.7685	18.7781	—

As Fig. 4 tells, the raw low-light image can indeed be enhanced by the traditional enhancement algorithm, but the shortcomings are obvious. The HE and gamma transform based enhancement are apt to the overly enhanced brightness. Although CLAHE generates better contrast improvement, it also demonstrates amplified noise as same as HE and gamma transform do, which leads to extremely low PSNR. After applying denoising post-processing as shown in Fig. 5, the PSNR can be improved as given in Table 2. To be specific, the Non-Local Means (NLM) denoising algorithm can greatly improve the image quality and remove most of the noise compared with the median filtering. However, it is still difficult to completely remove the extremely low-light image noise. In comparison, by using the neural network based enhancement algorithm proposed in this paper, the contrast and brightness of the low-light image can be improved, and the noise is also effectively suppressed. The PSNR value is higher and has a better visual effect.

4   Conclusion

When a traditional image enhancement method is applied to process a low-light image, the noise is significantly amplified, resulting in a low signal-to-noise ratio after the enhancement. Therefore, denoising processing must be further implemented. The end-to-end fully convolutional neural network is proposed in this paper unifies the low-light image enhancement and denoising problem in a framework. It can achieve very good enhancement and denoising effect for low-light image. Applying such method to the field of low-light remote sensing will greatly improve the application efficiency of low-light data, and has great application prospects.

References

1. Zhe, Bai, et al. The application of auto-gated power supply in ICCD camera. 2013.
2. Bai, Zhe, et al. "A study on the application of ICCD in low light level remote sensing."International Symposium on Photoelectronic Detection and Imaging 2013: Low-Light-Level Technology and Applications International Society for Optics and Photonics, 2013:89121G.
3. Guan-Zhang, L. I., W. S. Luo, and L. I. Pei. "Color Image Enhancement Based on Visual Characteristics of Human Eyes." Opto-Electronic Engineering 36.11(2009):92–95.
4. Lee, Eunsung, et al. "Contrast Enhancement Using Dominant Brightness Level Analysis and Adaptive Intensity Transformation for Remote Sensing Images." *IEEE Geoscience & Remote Sensing Letters* 10.1(2012):62–66.
5. Jang, Jae Ho, S. D. Kim, and J. B. Ra. "Enhancement of Optical Remote Sensing Images by Subband-Decomposed Multiscale Retinex With Hybrid Intensity Transfer Function." IEEE Geoscience & Remote Sensing Letters 8.5(2011):983–987.
6. Jaspers, Cornelis A. M. "Histogram equalization." US, US6741736. 2004.
7. Pizer, Stephen M., et al. "Adaptive histogram equalization and its variations." Computer Vision Graphics & Image Processing 39.3(1987):355–368.
8. Li, Zhang. "Contrast Limited Adaptive Histogram Equalization." Computer Knowledge & Technology (2010).

9. Farid, H. "Blind inverse gamma correction." IEEE Transactions on Image Processing A Publication of the IEEE Signal Processing Society 10.10(2001):1428.
10. Li, Yun. "ON ALGORITHM OF IMAGE CONTRAST ENHANCEMENT BASED ON WAVELET TRANSFORMATION." Computer Applications & Software (2008).
11. Loza, A, D. Bull, and A. Achim. "Automatic contrast enhancement of low-light images based on local statistics of wavelet coefficients." Digital Signal Processing 23.6(2013): 1856–1866.
12. Zhan Bichao, Wu Yiquan, and Ji Shouxin. "Infrared Image Enhancement Method Based on Stationary Wavelet Transformation and Retinex." Acta Optica Sinica 30.10(2010): 2788–2793.
13. Park, Seonhee, et al. "Low-light image enhancement using variational optimization-based retinex model." IEEE Transactions on Consumer Electronics 63.2(2017):178–184.
14. Kil, Tae Ho, H. L. Sang, and N. I. Cho. "A dehazing algorithm using dark channel prior and contrast enhancement." IEEE International Conference on Acoustics, Speech and Signal Processing IEEE, 2013:2484–2487.
15. He, Kaiming, J. Sun, and X. Tang. "Single Image Haze Removal Using Dark Channel Prior." IEEE Transactions on Pattern Analysis & Machine Intelligence 33.12(2011): 2341–2353.
16. Zhe, L. V., et al. "Region-adaptive Median Filter." Journal of System Simulation 19.23 (2007):5411–5414.
17. Bai, Lianfa, and B. Zhang. "Theory and experiment study on low light level image by median filter and mode filter."Journal of Nanjing University Ofence & Technology (1995).
18. Wang, Jie, et al. "Frequency domain reciprocal—Gaussian cascade low-pass filtering denoising method of image." Application Research of Computers 29.12(2012):4776–4778.
19. Danielyan, A, et al. "Cross-color BM3D filtering of noisy raw data." International Workshop on Local and Non-Local Approximation in Image Processing IEEE, 2010:125–129.
20. Harmeling, S., C. J. Schuler, and H. C. Burger. "Image denoising: Can plain neural networks compete with BM3D?." IEEE Conference on Computer Vision and Pattern Recognition IEEE Computer Society, 2012:2392–2399.
21. Dabov, K., et al. "Image Denoising by Sparse 3-D Transform-Domain Collaborative Filtering." IEEE Transactions on Image Processing 16.8(2007):2080–2095.
22. Zhang, Kai, et al. "Beyond a Gaussian Denoiser: Residual Learning of Deep CNN for Image Denoising." IEEE Transactions on Image Processing A Publication of the IEEE Signal Processing Society 26.7(2016):3142–3155.
23. Jain, Viren, and H. S. Seung. "Natural image denoising with convolutional networks." International Conference on Neural Information Processing Systems Curran Associates Inc. 2008:769–776.
24. Ronneberger, Olaf, P. Fischer, and T. Brox. "U-Net: Convolutional Networks for Biomedical Image Segmentation." 9351(2015):234–241.
25. Chen, Chen, et al. "Learning to See in the Dark." (2018).
26. Kingma D, Ba J. Adam: A Method for Stochastic Optimization[J]. Computer Science, 2014.
27. Huynh-Thu, Q., and M. Ghanbari. "Scope of validity of PSNR in image/video quality assessment." Electronics Letters 44.13(2008):800–801.

Imaging Algorithm of Arc Array MIMO-SAR Based on Keystone Transform

Pingping Huang[1,2], Xin Du[1,2(✉)], Wei Xu[1,2], and Weixian Tan[1,2]

[1] College of Information Engineering, Inner Mongolia University of Technology, Hohhot, Inner Mongolia 010051, China
duxinedu@163.com
[2] Inner Mongolia Key Laboratory of Radar Technology and Application, Inner Mongolia University of Technology, Hohhot, Inner Mongolia 010051, China

Abstract. Arc array Multiple-Input Multiple-Output Synthetic Aperture Radar (MIMO-SAR) is a novel array microwave imaging mode for wide-area observation. Due to the special characteristics of azimuthal sampling at equal angular intervals, the existing imaging algorithms of linear array are no longer suitable for this mode. As for this issue, a new arc array MIMO-SAR microwave imaging algorithm based on keystone transform is presented. This method corrects the movement of the target cross range unit by eliminating the coupling term between the azimuth and the range. The simulation results of point targets validate the presented imaging approach.

Keywords: Arc array · Keystone transform · Range cell migration correction (RCMC) · Imaging algorithm

1 Introduction

In recent years, a new MIMO-SAR imaging mode based on arc array antenna [1–4] has been proposed to against the deficiency of linear array in wide-area observation. Compared with the linear array imaging modes [5–9], this new imaging mode achieves a large observation angle. Multiple antenna array elements are arranged along the circular arc to form an arc synthetic aperture, so that the array resolution is achieved through synthetic aperture theory. Meanwhile, the high-speed microwave switch network, which is used to control the transmission and reception of signals, achieves multiple-input multiple-output mechanism.

Since the azimuth sampling of the arc array imaging mode is different from the linear array, the existing imaging methods for linear array imaging modes are no longer apply to this arc array imaging mode. Moreover, it is difficult to obtain the signal model in the Range-Doppler domain [10], so it is a hard task to take the conventional method to correct the range cell migration. This paper studies an arc array MIMO-SAR imaging method based on Keystone transform [11–14], which corrects the movement of the target cross range unit by eliminating the coupling term of the azimuth and the range in the range frequency domain. Then, the focused image is obtained by azimuth matching filtering.

© Springer Nature Singapore Pte Ltd. 2019
L. Wang et al. (eds.), *Proceedings of the 5th China High Resolution Earth Observation Conference (CHREOC 2018)*, Lecture Notes in Electrical Engineering 552,
https://doi.org/10.1007/978-981-13-6553-9_8

This paper is arranged as follows. In Sect. 2, the observation geometry and basic principle of arc array MIMO-SAR is introduced. At the same time, the echo signal model of arc MIMO-SAR is analyzed. In Sect. 3, the processing method of arc array MIMO-SAR imaging based on Keystone transform is studied. In Sect. 4, the imaging simulation is carried out, and the effectiveness and feasibility of the presented imaging method is validated by the simulation results of point targets and distributed target. Finally, the summaries are made.

2 Working Principle and Signal Model

2.1 Imaging Geometry

The imaging geometry of the arc array MIMO-SAR is shown in Fig. 1. The range resolution is realized by transmitting the bandwidth signal; meanwhile, the azimuth resolution is achieved through the arc synthetic aperture that is formed by multiple antenna array elements arranged along the circular arc.

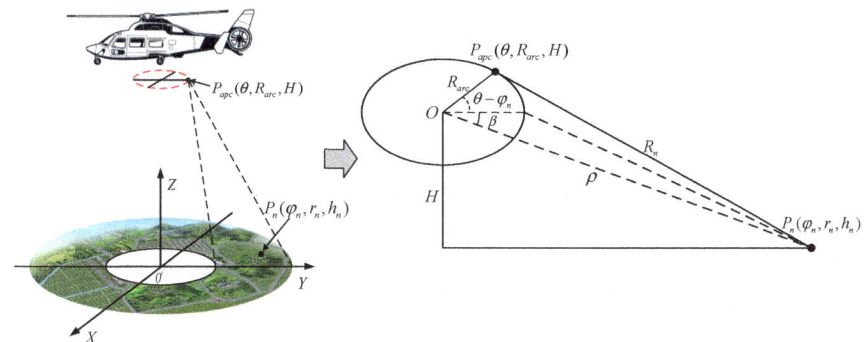

Fig. 1. Arc array MIMO-SAR imaging geometry

Figure 1 shows the imaging geometry model of arc array MIMO-SAR. O is the center of arc array, (θ, R_{arc}, H) is the position coordinate of arc array equivalent sampling point P_{apc}, θ is the horizontal angle of the equivalent sampling point, R_{arc} is the horizontal radius of the arc where arranging antenna element, H is the height of arc antenna array. (φ_n, r_n, h_n) is position coordinate of the point target P_n in the observed scene, φ_n is the horizontal angle of the point target, r_n is the horizontal radius from point target to the center of arc array, and h_n is height of the point target. R_n represents the instantaneous distance from equivalent sampling point P_{apc} to the point target P_n, ρ is the distance from point target P_n to arc array center O. β is the angle between the arc antenna array plane and OP_n.

The arc array MIMO-SAR imaging adopts frequency-modulation continuous-wave (FMCW) radar system that has many advantages such as small size and low weight. As shown in Fig. 2, the system is bistatic antenna configuration that its transmitter and receiver systems are separated [3]. The horizontal angle spacing between the equivalent center of arbitrary neighborhood antenna elements are all $\Delta\theta_{Int}$, the high-speed microwave switch network is used to control the transmission and reception.

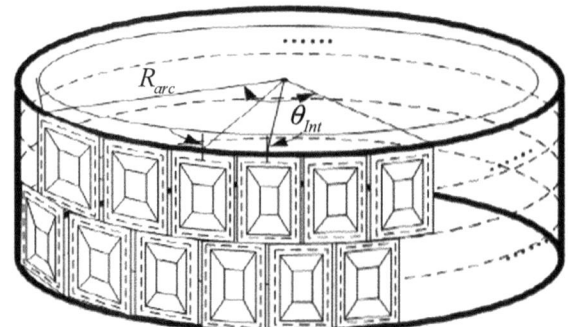

Fig. 2. The antenna structure of arc array MIMO-SAR imaging

2.2 Signal Model

As described above,the arc array MIMO-SAR imaging system adopts FMCW, so that its transmitted signal can be expressed as

$$S_{tr}(t) = rect\left(\frac{t}{T_r}\right) \exp\left[j2\pi\left(f_c t + \frac{1}{2}K_r t^2\right)\right] \tag{1}$$

where f_c is the radar carrier frequency, t is time variable of range, K_r is chirp rate of signals, and T_r represents the signal duration. The echo signal S_{re} that received from point target P_n at the equivalent sampling point P_{apc} can be expressed as

$$S_{re}(t, \theta) = \sigma_n(\theta_n, r_n, h_n) rect\left(\frac{t - 2R_n/c}{T_r}\right) rect\left(\frac{\theta - \varphi_n}{\theta_a}\right)$$
$$\cdot \exp\left\{j2\pi\left[f_c\left(t - \frac{2R_n}{c}\right) + \frac{1}{2}K_r\left(t - \frac{2R_n}{c}\right)^2\right]\right\} \tag{2}$$

where $\sigma_n(\theta_n, r_n, h_n)$ stands for the backscattering coefficient, θ_a is the beam width in the horizontal direction. The demodulated signal obtained by mixing the echo signal and the transmitted signal can be expressed as

$$S_{re\_IF}(t, \theta) = \sigma_n(\theta_n, r_n, h_n) rect\left(\frac{t - 2R_n/c}{T_r}\right) rect\left(\frac{\theta - \varphi_n}{\theta_a}\right)$$
$$\cdot \exp\left\{j2\pi\left[\frac{2R_n}{c}f_c + \frac{2R_n}{c}K_r t - \frac{1}{2}K_r\left(\frac{2R_n}{c}\right)^2\right]\right\} \tag{3}$$

The third term $j\pi K_r(2R_n/c)^2$ is the residual video phase (RVP) and it needs to be eliminated. The echo signal after RVP elimination and range IFFT can be described as

$$S_{\text{re\_rvp}}(t, \theta) = \sigma_n(\theta_n, r_n, h_n)p_r\left(t - \frac{2R_n}{c}\right)\text{rect}\left(\frac{\theta - \varphi_n}{\theta_a}\right)\exp\left[j\frac{4\pi f_c R_n}{c}\right] \quad (4)$$

where $p_r(\cdot)$ represents the envelope function after range compression, which is the sinc function usually. It can be observed in the Eq. (4) that R_n varies with the array changes, so the envelope of the range will move around.

3 Imaging Algorithm

3.1 Keystone Transform

Compared with linear array SAR imaging, the arc array MIMO-SAR imaging can achieve large-angle observation for the scene around the platform. The arc array MIMO-SAR imaging is based on equal angle-interval sampling rather than linear array equidistant interval sampling, so the existing linear array imaging algorithms are no longer suitable for this mode.

The Keystone transform needs to be performed in the range frequency domain. After a range FFT for the range-compressed signal shown in Eq. (4), the result can be expressed as

$$
\begin{aligned}
S_{\text{re\_rvp}}(f, \theta) = \sigma_n(\theta_n, r_n, h_n)\text{rect}\left(\frac{t - 2R_n/c}{T_r}\right)&\text{rect}\left(\frac{\theta - \varphi_n}{\theta_a}\right)\\
&\cdot \exp\left[j\frac{4\pi}{c}(f_c + f_r)R_n\right]
\end{aligned}
\quad (5)
$$

where $f_r = K_r t$. It can be seen from Fig. 1 that R_n can be written as

$$R_n = \sqrt{\rho^2 + R_{\text{arc}}^2 - 2\rho R_{\text{arc}}\cos\beta\cos(\theta - \varphi_n)} \quad (6)$$

In fact, the arc array radius $R_{\text{arc}} \ll \rho$, and therefore, (6) can be approximated as

$$R_n \approx \rho - R_{\text{arc}}\cos\beta\cos(\theta - \varphi_n) \quad (7)$$

Equation (5) can be rewritten as

$$
\begin{aligned}
S_{\text{re\_rvp}}(f, \theta) \dot{=} \sigma_n \text{rect}\left(\frac{t - 2R_n/c}{T_r}\right)&\text{rect}\left(\frac{\theta - \varphi_n}{\theta_a}\right)\exp\left[j\frac{4\pi}{c}(f_c + f_r)\rho\right]\\
&\cdot \exp\left[-j\frac{4\pi}{c}(f_c + f_r)R_{\text{arc}}\cos\beta\cos(\theta - \varphi_n)\right]
\end{aligned}
\quad (8)
$$

In order to facilitate the analysis, Eq. (8) can be rearranged to

$$S_{re\_rvp}(f, \theta) = \sigma_n \text{rect}\left(\frac{t - 2R_n/c}{T_r}\right) \text{rect}\left(\frac{\theta - \varphi_n}{\theta_a}\right) \exp\left[j\frac{4\pi}{c}(f_c + f_r)\rho\right]$$
$$\cdot \exp\left[-j\frac{4\pi}{c}f_c R_{arc} \cos \beta \cos(\theta - \varphi_n)\right] \qquad (9)$$
$$\cdot \exp\left[-j\frac{4\pi}{c}K_r t R_{arc} \cos \beta \cos(\theta - \varphi_n)\right]$$

For Eq. (9), the third index term\exp$[-j4\pi K_r t R_{arc} \cos \beta \cos(\theta - \varphi_n)/c]$ is the main cause of range envelope migration. There is a coupling of the range time and the array angle. Eliminating the coupling that can correct the movement of the target cross range unit.

According to the relationship of Keystone transform, redefine a virtual azimuth sampling α, and the relationship between θ and α is

$$\cos(\theta - \varphi_n) = \frac{f_c}{f_c + f_r} \cos(\alpha - \varphi_n) \qquad (10)$$

Substituting Eq. (10) into Eq. (8) and IFFT, we then get

$$S_{re\_rvp}(t, \alpha) = \sigma_n(\theta_n, r_n, h_n)p_r\left[t - \frac{2\rho}{c}\right]\text{rect}\left(\frac{\alpha - \varphi_n}{\theta_a}\right)\exp\left[j\frac{4\pi f_c}{c}\rho\right]$$
$$\cdot \exp\left[-j\frac{4\pi f_c}{c}R_{arc} \cos \beta \cos(\alpha - \varphi_n)\right] \qquad (11)$$

Equation (11) is the time domain echo signal after range migration correction, and coupling term is eliminated through Keystone transform. The range envelope no longer changes with the azimuth angle, and the echoes of different range units are corrected to the same range unit.

3.2   Azimuth Process

The azimuth matched filter needs to be performed in the Range-Doppler domain, so have an azimuth FFT for Eq. (11). In this algorithm, the azimuth matched filter is obtained by the fast convolution that can be achieved by conjugate operation and fast Fourier transform. The kernel function for the convolution is

$$h_{az}(\alpha, \varphi_n) = \exp\left[-j\frac{4\pi f_c}{c}R_{arc} \cos \beta \cos(\alpha - \varphi_n)\right] \qquad (12)$$

The azimuth matched filter is calculated as [10]

$$H_{az}(\alpha, \varphi_n) = \{FFT[h_{az}(\alpha, \varphi_n)]\}^* \qquad (13)$$

where FFT represents the fast Fourier transform, symbol $*$ denotes the complex conjugate operation.

After matched filter and azimuth inverse Fourier transform, the signal is compressed as

$$S_{\text{re\_rvp}}(t, \alpha) = \sigma_n(\theta_n, r_n, h_n) p_r \left[t - \frac{2\rho}{c} \right] p_a[\alpha - \varphi_n] \exp\left[j\frac{4\pi f_c}{c} \rho \right] \tag{14}$$

Here, $p_a(\cdot)$ is the envelope function in the azimuth direction, which is the sinc function usually.

The acquired echo data is stored in a polar coordinate system and we can transform it from polar coordinate to Cartesian coordinate by the following formula:

$$\begin{cases} x_n = \rho \cos(\alpha - \varphi_n) \\ y_n = \rho \sin(\alpha - \varphi_n) \end{cases} \tag{15}$$

where (x_n, y_n) is the position of the target in the Cartesian coordinate system. The imaging process can be summarized as shown in Fig. 3.

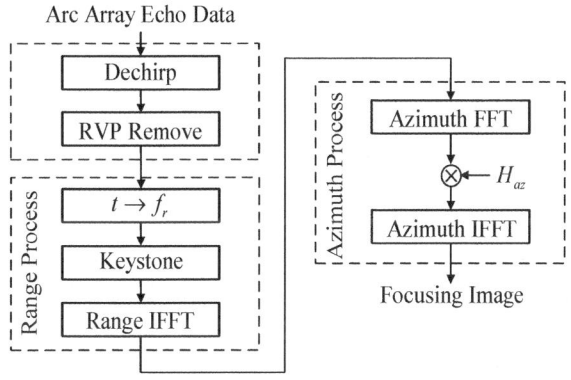

Fig. 3. Procedure of arc array MIMO-SAR imaging algorithm

The resolution of the arc array MIMO-SAR imaging mainly depends on the bandwidth of the transmitted signal, the radius of the arc array, and the distance from target to antenna (Fig. 4). The range resolution [3] can be expressed as

$$\rho_r = \frac{c}{2B_r(1 - \tau_n/T_r)\cos\beta} \tag{16}$$

where B_r is the signal bandwidth, τ_n is delay time of the echo, and T_r represents the signal duration.

Azimuth instantaneous frequency f_θ is

$$
\begin{aligned}
f_\theta &= \frac{\partial}{\partial \theta} \{\arg[S_{re\_IF}(t, \theta)]\} \\
&= \frac{4\pi f}{c} \frac{R_{arc}\rho \sin(\theta - \theta_n)}{\sqrt{R_{arc}^2 + \rho^2 - 2R_{arc}\rho \cos(\theta - \theta_n) + (H - h_n)^2}}
\end{aligned}
\tag{17}
$$

Here, $f = f_c + K_r t$, azimuth resolution is [3]

$$
\rho_\theta = \frac{2\pi}{\max\{f_\theta\} - \min\{f_\theta\}} \approx \frac{\lambda\sqrt{R_{arc}^2 + \rho^2 - 2R_{arc}\rho \cos(\theta_a/2) + (H - h_n)^2}}{4R_{arc}\rho \sin(\theta_a/2)}
\tag{18}
$$

where λ is wavelength of the signal and θ_a is the horizontal beam width.

It can be seen from Eq. (18) that the azimuth resolution is related to the height of platform and the distance from the target to the center of the arc array antenna (Fig. 5).

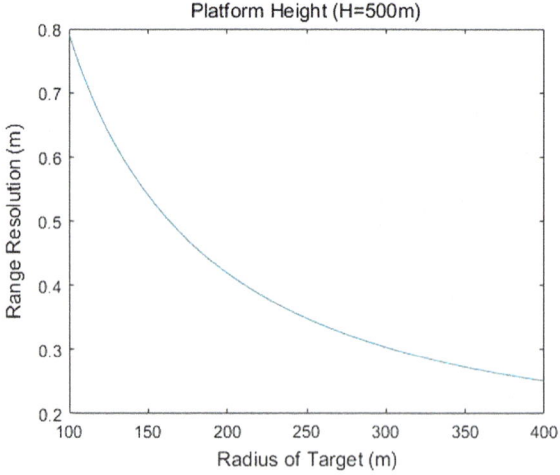

Fig. 4. Range resolution varying with target range at the same altitude

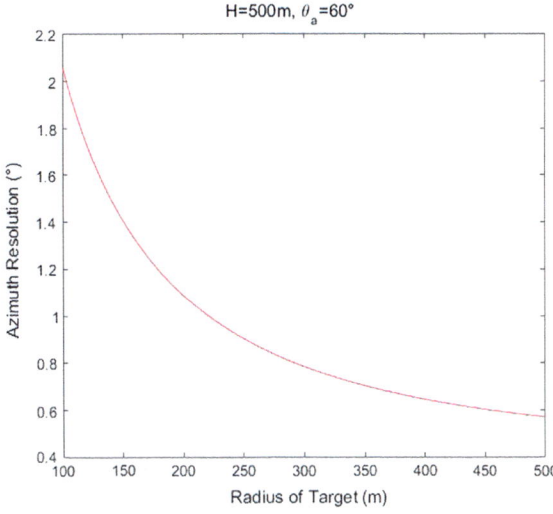

Fig. 5. Azimuth resolution varying with target range at the same altitude

4 Simulation Analysis

In order to validate the effectiveness and feasibility of the imaging method in this paper, a simulation experiment of the point targets was carried out. The simulation parameters of the arc array MIMO-SAR system are shown in Table 1.

Table 1. Simulation parameters

Symbol	Definition	Value
f_c	Carrier frequency	35.5 GHz
B_r	Signal bandwidth	1000 MHz
T_r	Sweep time	0.1 ms
H	Platform height	500 m
R_{arc}	Arc array radius	0.6 m
θ_{inc}	Incident angle	25°
θ_e	Pitch beamwidth (−3 dB)	30°
θ_a	Array beamwidth (−3 dB)	60°

As shown in Fig. 6, the scene of target distribution is divided into four regions, and a total of 10 targets are distributed around the platform. The radius of the three targets P1, P2, and P3 in the area C are 300, 305, and 305 m. Meanwhile, the angles are 0°, 10°, and −10°.

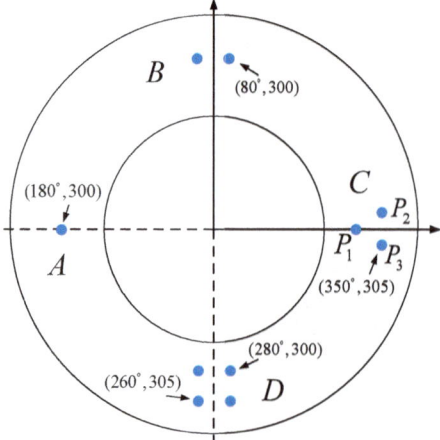

Fig. 6. The positions of targets

Fig. 7. The imaging results of targets

The imaging algorithm in this paper is used to process the echo data and the focus imaging is given in Fig. 7. Figure 8 gives the contours of the three targets P1, P2, and P3 in the area C, and it can be seen that the three targets get a good focus.

The specific imaging performance of targets P1, P2, and P3 are given in Table 2. Where PSLR represents peak sidelobe ratio and ISLR is the integrated sidelobe ratio. It can be observed from Eqs. (16) and (18) that the range and azimuth resolutions change with the variations of target location. The performance of imaging results is consistent with the theoretical analysis, which validates the effectiveness and feasibility of the proposed algorithm.

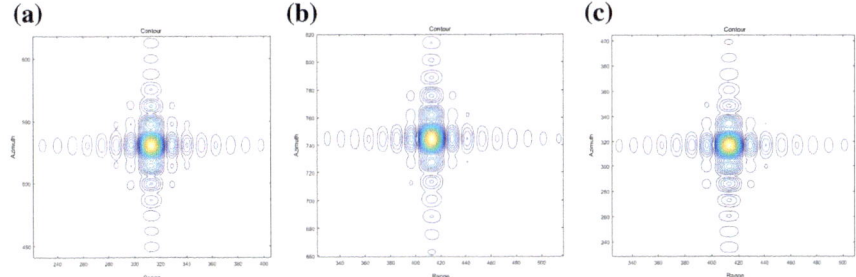

Fig. 8. Contour map. **a** Target P1; **b** Target P2; **c** Target P3

Table 2. The analysis for imaging performance

Target	Range				Azimuth			
	IRW(m)		PSLR	ISLR	IRW(°)		PSLR	ISLR
	Theory	Actual	(dB)	(dB)	Theory	Actual	(dB)	(dB)
P1	0.292	0.301	−13.41	−9.48	0.784	0.791	−12.66	−8.85
P2	0.289	0.297	−13.64	−9.73	0.775	0.779	−12.55	−8.61
P3	0.289	0.295	−13.58	−9.62	0.775	0.780	−12.68	−8.74

5 Conclusion

As a special array imaging mode with circular trace scanning, the arc array MIMO-SAR greatly increases the angle that observation the scene around platform. The echo signal special characteristics of arc array MIMO-SAR imaging are analyzed and the imaging algorithm based on Keystone transform is presented in this paper. The algorithm corrects the movement of the target cross range unit through Keystone transform. The simulation results of the target validate the imaging method in this paper.

References

1. Huang Pingping. Research on Imaging Algorithm with MIMO-SAR based on Arc Antenna Array [C]. High Resolution Earth Observation Conference, 2014.
2. Huang Pingping, Tan Weixian, Su Yin. MIMO-SAR imaging technology for helicopter-borne based on ARC antenna array [C]. IEEE Geoscience and Remote Sensing Symposium, 2015, 1801–1804.
3. Huang Pingping, Tan Weixian, Su Yin, et al. Huang Pingping, Tan Weixian, Su Yin, et al. Research on Helicopter-borne MIMO Microwave Imaging Technology Based on Arc Antenna Array [J]. Journal of Radars, 2015, 4(1):11–19.
4. Huang Pingping, Tan Weixian, Hong Wen. Microwave signal transceiving system, method and imaging system for MIMO-SAR imaging [P]. 2015.

5. Huang Pingping, Deng Yunkai, Xu Wei. Research on Multiple-Input and Multiple-Output SAR Imaging Based on CS Algorithm [J]. Journal of University of Electronic Science and Technology of China, 2012. 41(2): 222–226.
6. Huang Pingping, Li Hengchao, Feng Fan, et al. The Waveform Analysis and Signal Processing for Space-borne MIMO-SAR [C]. European Conference on Synthetic Aperture Radar, 2012, 579–582.
7. Tan Weixian. Study on Theory and Algorithms for Three-Dimensional Synthetic Aperture Radar Imaging [D]. Beijing: Graduate University of Chinese Academy of Sciences, 2009.
8. Shi Jun. Research on Principles and Imaging Techniques of Bistatic SAR and linear array SAR [D]. Chengdu: University of Electronic Science and Technology of China, 2009.
9. Du Lei. Study on Model, Algorithm and Experiment for Downward-Looking Synthetic Aperture Radar Three-Dimensional Imaging Based on Linear Array Antennas [D]. Beijing: Graduate University of Chinese Academy of Sciences, 2010.
10. Luo Yunhua, Song Hongjun, Wang Yu, et al. Signal Processing of Arc FMCW SAR [C] APSAR 2013, 2013.
11. Wang Shuyan, Su Zhigang, Wu Renbiao. A Modified RD Imaging Method Using the Keystone Transform and Rank-deficient Capon for High Squint SAR [J]. Modern Radar, 2011, 33(9):25–28.
12. Shao Hongxu, Liao Chen, Ji Xiaoyu, et al. Application of Keystone Transform for Millimeter-Wave Radar Signal Processing [J]. Journal of Microwaves, 2017.
13. [13] Wang Juan, Zhao Yongbo. Research on Implementation of Keystone Transform [J]. Fire Control Radar Technology, 2011, 40(1):45–51.
14. Zeng Tao, Hu Cheng, Mao Cong, et al. A rapid imaging method for ground-based synthetic aperture radar based on Keystone transform [P]. 2015.

Performance Analysis of Vehicle-Borne Dual-Band Dual-Polarization Weather Radar

Weixian Tan[1,2], Xiangtian Zheng[1(✉)], Pingping Huang[1,2], Wei Xu[1,2], and Kuoye Han[3]

[1] College of Information Engineering, Inner Mongolia University of Technology, Hohhot, Inner Mongolia 010051, China
xtzheng1020@163.com
[2] Inner Mongolia Key Laboratory of Radar Technology and Application, Inner Mongolia University of Technology, Hohhot, Inner Mongolia 010051, China
[3] Electronics Technology Group Corporation, Information Science Academy, Beijing, P. R. China 100098, China

Abstract. Dual-band and dual-polarization are the trends of development of weather radar, particularly for real-time clouds and precipitation observation applications. In order to strengthen the monitoring of the spatial structure of the precipitation process and increase the possibility of artificial intervention in the weather, combining the advantages of Ka-band and X-band are used to measure clouds and light rain and other meteorological targets, respectively, this paper studies a vehicle-borne dual-band dual-polarization weather radar system. Based on the vehicle-borne, the system adopts Ka/X dual-bands for joint detection of meteorological targets of clouds, fog and precipitation, and it features single-band-independent detection and dual-band joint detection to enhance observation of clouds and light rain. The radar is equipped with a blind compensation pulse for each pulse period, which is used to compensate for the blind area that cannot be detected under normal conditions, and improves the radar range. First, this paper introduces in detail the system principle and configuration of the dual-band dual-polarization weather radar. Second, the operation parameters and performance are further analysed.

Keywords: Weather radar · Dual-band · Dual-polarization · Vehicle-borne · Cloud and precipitation observation

1 Introduction

In order to obtain accurate information in atmospheric observations, it is necessary to measure the radar echo of the rain and clouds. In general, weather radars from S-band to Ku-band can all observe rainfall well, while millimetre-wave radars can detect small particles smaller than the radar wavelength. Compared with conventional radar, dual-band weather radars can measure and describe multiple parameters of the drop-size distribution, and these parameters are widely used to quantitatively measuring rainfall and hydrometeor type and help to improve inversion precision of the intensity of clouds and rain.

© Springer Nature Singapore Pte Ltd. 2019
L. Wang et al. (eds.), *Proceedings of the 5th China High Resolution Earth Observation Conference (CHREOC 2018)*, Lecture Notes in Electrical Engineering 552,
https://doi.org/10.1007/978-981-13-6553-9_9

In the 1950s, scientists studied the backscattering of polarized waves by atmospheric particles [1, 2]. In the late 1970s, Seliga and Bringi [3] proposed the idea of dual-polarization radar and measured the precipitation intensity, and the results show that the precision of dual-polarization radar is higher than that of conventional radar. In the United States, WSR-88D (NEXRAD) Doppler weather radar network will start to get upgraded to dual-polarization capability [4] in 2010.

In addition to obtaining reflectivity factor, radial velocity and spectrum width of the meteorological target, dual-polarization weather radar can also obtain parameters, such as differential reflectivity, specific differential phase, correlation coefficient, linear depolarization ratio. These parameters can more accurately show droplet spectrum characteristics of meteorological particles. We introduces a vehicle-borne dual-band dual-polarization weather radar system. This system works in single-output and dual-input mode of operation [5–7]. The system combines millimetre and centimetre waves, larger precipitation particles can be observed in the X-band, and smaller meteorological particles, such as clouds and fog can be observed in the Ka-band. Meanwhile, the miniaturization and lightweight of this radar are easy to implement [8].

First, this paper introduces the system principle and the configuration of vehicle-borne dual-band dual-polarization weather radar. Second, through the analysis of radar system parameters and performance, indicating the system has a good performance and full functionality, and can be a good observation of the evolution of weather conditions.

2   System Principle and Configuration

The purpose of using vehicle-borne dual-band dual-polarization weather radar is to capture well weather changes. In practical applications, conventional radar can only detect meteorological parameters to a certain extent. In order to study the precipitation process and the internal structure of the cloud, radars need to be able to observe clouds and rain at the same time [9].

2.1   System Principle

As shown in Fig. 1a, the vehicle-borne dual-band dual-polarization weather radar operates in Ka-band and X-band, and this radar adopts all-solid-state, single-transmitting and double-receiving linear polarization, coherent pulse Doppler system. The system combines the advantages of the Ka-band in detecting clouds and X-bands in detecting rain, and uses the principle of backscattering of electromagnetic waves by atmospheric particles to obtain the dual-polarization parameters of the detected targets [10].

The observation geometry of weather radar is shown in Fig. 1b, where x-y is horizontal plane and z is vertical height direction. The system uses dual-polarization antennas, horizontal and vertical polarized electromagnetic waves are alternately transmitted through a dual-channel transmitter.

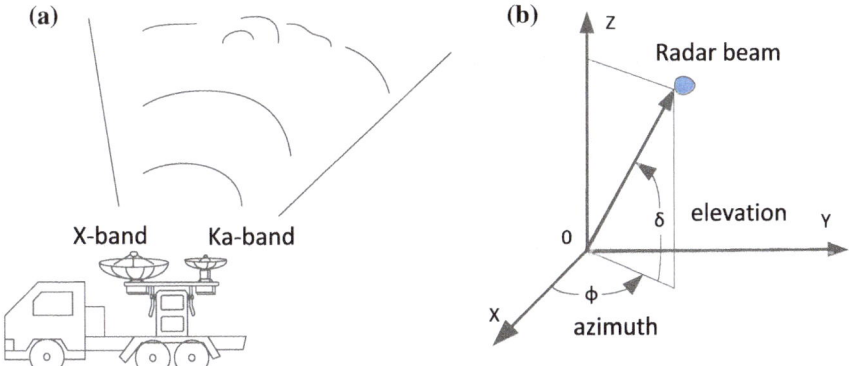

Fig. 1. Vehicle-borne dual-band dual-polarization weather radar system structure: **a** Weather radar schematic diagram, **b** observation geometry

The target scattering matrix S can be measured within two pulse times. The target scattering matrix S associates the incident electric field with the scattered electric field [11].

$$S = \begin{pmatrix} S_{hh} & S_{hv} \\ S_{vh} & S_{vv} \end{pmatrix} \tag{1}$$

The matrix element is S_{ij}, subscript i represents the polarization state of backscattering, subscript j represents the polarization state of the electric field at transmitting. Dual-polarization weather radar can also obtain polarization parameters of the meteorological target, including linear depolarization ratio, differential reflectivity, specific differential phase, correlation coefficient and so on. The processing of echo polarization information can obtain the second products inversion of meteorological data to further understand the state of internal particle forms and movements under meteorological conditions such as clouds, light rain and fog.

2.2   System Structure and Components

The structure diagram of vehicle-borne dual-band dual-polarization weather radar is shown in Fig. 2. The radar has two parabolic antennas, Ka-band antenna diameter $D_{km} = 0.6\,\text{m}$ and $f_{ka} = 35.5\,\text{GHz}$, it is used to observe clouds, X-band antenna diameter $D_x = 2.4\,\text{m}$ and $f_x = 9.375\,\text{GHz}$, is used to observe precipitation.

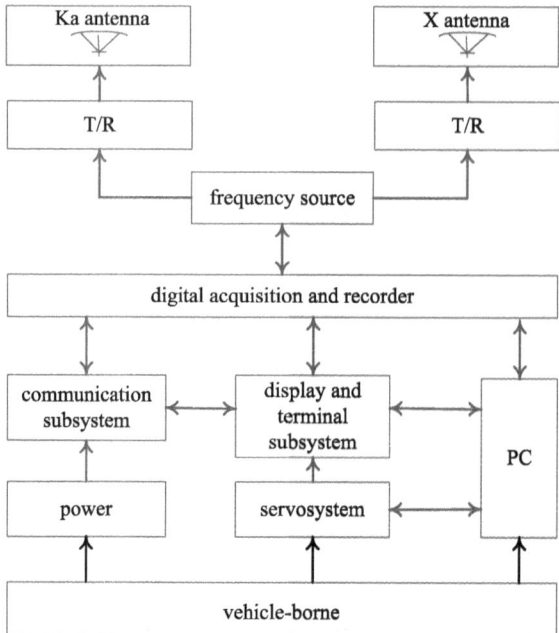

Fig. 2. Structure diagram

The system is mainly composed of radar equipment and vehicle-borne. The radar equipment includes antennas and TR module, calibration subsystem, servo system, digital acquisition and recorder system, display and terminal subsystem, communication subsystem. Amongst them, the antennas, TR module and calibration subsystem are independently equipped with Ka-band and X-band. The servo system, digital acquisition and recorder system, display and terminal subsystem, communication subsystem and vehicle platforms are common to Ka-band and X-band.

The high-power pulse signal produced by the frequency source enters the TR module. In the TR module, the transmitter divides the input signal into horizontal and vertical polarizations and radiate into the air through the antennas. The echo received by the receiver mainly comes from the backscattering of the meteorological target. The echo signal received by the antennas is divided into two signals of horizontal polarization and vertical polarization by a coupler, and then enter the respective receiver.

Digital acquisition and recorder system mainly include digital receiver and signal processor.

The calibration subsystem has built-in test equipment. It can automatically detect the parameters such as transmit power and receiving characteristic curve.

The display and terminal subsystem can provide real-time meteorological data such as location, echo power, radial velocity, spectrum width and polarization feature, it can ensure the effective inversion of the second products, and the data can also be processed remotely through the communication subsystem.

The servo system is mainly used to control the antenna azimuth and elevation scanning, and upload the fault information [12–14].

3   System Parameters and Performance Analysis

3.1   Weather Radar Working Mode

Combining the computational power of the signal processor with the complexity of the ultra-low sidelobe pulse compression algorithm, the pulse compression signal is selected as the detection signal of the radar system. Through the waveform design of the transmitting signal and channel delay equalization correction, it can effectively suppress the influence of strong reflection targets in the distance direction on the weak reflection target. In order to meet the requirements of different detection distances and detection height set low elevation modes and high elevation modes.

The entire radar mounted on vehicle-borne. After determining the area to be measured, a suitable position can be selected to observe the meteorological target. The radar has a variety of working methods, the fixed mode is the scanning mode that fixes the radar beam to a fixed point and keeps the position of azimuth and elevation angle unchanged. Plan Position Indicator (PPI) scan is a scanning method in which the elevation of the radar is fixed and the azimuth angle is rotated, azimuth scanning range is 0–360°, and the elevation range is −2–182°. Radar Height Indicator (RHI) enables radar to complete profile analysis of a certain detection area, in this mode, the azimuth is fixed for elevation scanning. VOL (volume scan) is composed of multiple PPI scans of different height levels, starting from a low elevation and completing the PPI scan layer by layer, VOL has a great use for studying the distribution of clouds in the entire space.

When the radar works at low elevation, the elevation ranges from 0 to 10° to detect distant weather targets. In another case, when the radar is working at high elevation, the elevation ranges from 10 to 90°, which is used to achieve high elevation close range meteorological target detection [15].

3.2   *Radar* Equation in Meteorology

The radar equation in meteorology [16, 17] is

$$P_r = \frac{\pi^3}{1024(ln2)} \frac{P_t G^2 \theta \varphi \tau c}{\lambda^2 L} \frac{1}{R^2 10^{K_r R}} |K|^2 \psi Z \tag{2}$$

Known by formula (2):

$$dBZ = 10 \lg \frac{2.78 \lambda^2}{P_t \theta \varphi \tau} + P_r + L + 20 \lg R - 2G - 10 \lg \psi + 160 \tag{3}$$

where R is range, P_r is received power, P_t is transmitted power, G is antenna gain function, θ is antenna azimuth beamwidth, φ is antenna elevation beamwidth, τ is radar pulsewidth, λ is wavelength, L is radar system loss, K_r is atmospheric attenuation, $|K|^2$ is dielectric constant, ψ is filling coefficient, Z is radar reflectivity factor, K is Boltzmann constant, B is bandwidth, F is radar noise figure, N_{fft} is FFT points, P_{rmin} is minimum detectable signal in the system.

$$P_{r\min} = \frac{KTBF}{N_{\text{fft}}}\text{SNR} \tag{4}$$

The system does not consider atmospheric attenuation, and uses the parameters in Table 1 to calculate the minimum radar reflectivity factor under different spatial distributions. The results are shown in Figs. 3 and 4.

Table 1. Radar parameter

Parameter	X	Ka
P_t	450 W	200 W
G	43 dB	43 dB
θ	1°	1°
φ	1°	1°
τ	60 μs, 20 μs	60 μs, 20 μs
c	3×10^8 m/s	3×10^8 m/s
λ	0.03200 m	0.00854 m
L	3 dB	5 dB
$|K|^2$	0.9	0.9
ψ	1	1
B	5 MHz	5 MHz
F	3.0	5.5
N_{fft}	256	256
SNR	13 dB	13 dB

3.3   System Waveform Design

In order to meet the requirements of different detection distances and detection heights, the radar sets two working modes, low mode and high mode, and the working waveforms used in different working modes as shown in Fig. 5. Three pulses are transmitted during each pulse recurrent time. Where τ_1 is the test pulse, which is used to test the gain and noise level of the receiving channel in real time. τ_2 is blind compensation pulse, which is used to compensate for the blind area that cannot be detected by a normal pulse. τ_3 is a normal working pulse and is used to generate transmit power for target detection.

When the radar is working in the high mode, the 2 μs narrow pulse and the 20 μs chirp signal are alternately transmitted, which is used for detecting the close range meteorological target, and has the characteristics of high pulse repetition frequency and narrow transmission pulse width. When the radar is working in the low mode, the 6 μs narrow pulse and the 60 μs chirp signal are alternately transmitted, which is used for detecting the long-range meteorological target. In the case of taking into account the close range blind area and detection capability, the pulse of 6 μs is used as the blinding, and the blind distance after blind complement is about 0.9 km. As can be seen from Figs. 3 and 4, after blind complement, the radar detection capability is increased by about 10 dB in the range of 3–9 km.

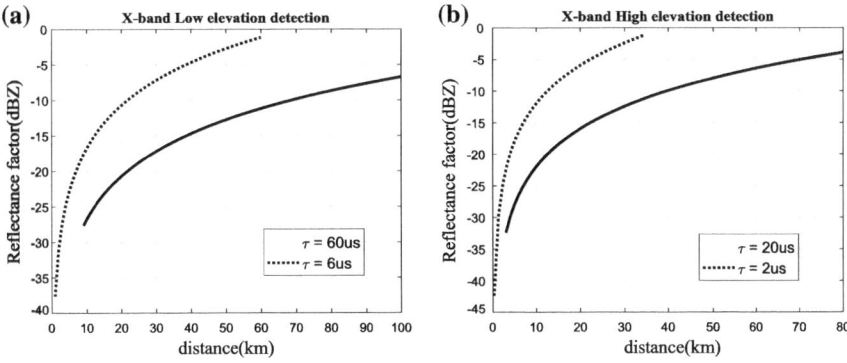

Fig. 3. X-band detection capability, **a** is X-band low elevation detection, **b** X-band high elevation detection

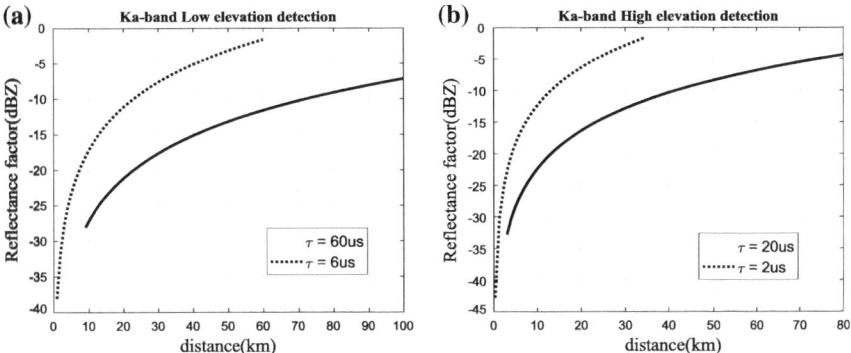

Fig. 4. Ka-band detection capability, **a** is Ka-band low elevation detection, **b** is Ka-band high elevation detection

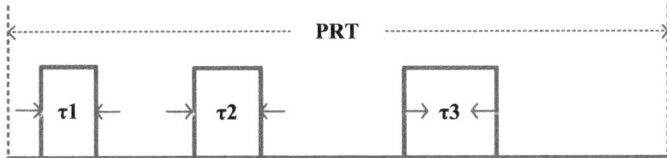

Fig. 5. Radar working waveform in pulse recurrent time

3.4 System Sensitivity Analysis

Known by Formula (4) radar system receiving sensitivity formula is:

$$P_{r\min} = 90 + 10\lg(\text{KTB}) + F + \text{SNR} - 10\lg(N_{\text{fft}}) \tag{5}$$

where in Ka-band F is 5.5 dB and B is 5 MHz, the Ka-band receiver limit system sensitivity is -112.568 dBm. In X-band, F is 3.0 dB and B is 5 MHz, the X-band receiver limit system sensitivity is -115.068 dBm [18, 19].

4 Conclusion

The vehicle-borne dual-band dual-polarization weather radar uses two bands, Ka-band and X-band, for better observation of important meteorological targets such as clouds and rain. This paper introduces the system principle and radar configuration as well as the working method of the radar, and analyzes the performance of the radar in detail. This system takes advantage of the Ka-band observation cloud and the X-band observation rain, which can increase the working distance of the radar, and can monitor the weather conditions in real time, making the possibility of artificial intervention of the weather increase.

Acknowledgements. This work is supported by the National Natural Science Foundation of China under Grant Nos. 61631011 and 61461040 and 61661043, and Natural Science Foundation of Inner Mongolia Autonomous Region under Grant No. 2016MS0606.

References

1. Atlas D, Kerker M, Hitschfeld W.: 'Scattering and attenuation by non-spherical atmospheric particles'. Journal of Atmospheric & Terrestrial Physics, 1953, 3(2), PP 108–119
2. Gent H, Hunter I.M, Robinson N.P.: 'Polarization of radar echoes, including aircraft, precipitation and terrain'. Electrical Engineers Proceedings of the Institution of, 1963, 110 (12). PP 2139–2148
3. Seliga T.A, Bringi V.N.: 'Differential reflectivity and differential phase shift: Applications in radar meteorology'. Radio Science, 2016, 13(2), PP 271–275
4. Istok M.J, Fresch M, Jing Z, et al.: '15.5 WSR-88D DUAL POLARIZATION INITIAL OPERATIONAL CAPABILITIES'. Roc.noaa.gov, 2009
5. Battan, L.J.: 'Radar Observation of the Atmosphere' (University of Chicago Press, 1973.)
6. Doviak, R.J., Zrnic, D.S.: 'Doppler Radar and Weather Observation' (NEW YORK: Dover Pubns, 2006.)
7. Stiglitz, M.R., Blanchard, C.: 'Spaceborne Weather Radar', Norwood Ma Artech House P, 1991
8. Pozar, D.M., Schaubert, D.H., Targonski, S.D., et al.: 'A dual-band dual-polarized array for spaceborne SAR'. Antennas and Propagation Society International Symposium. IEEE, 1998, pp 2112–2115

9. Patyuchenko, A., Younis, M., Krieger, G., et al.: 'Compact X/Ka-band dual-polarization spaceborne digital beamforming Synthetic Aperture Radar'. Radar Symposium. IEEE, 2015, pp. 1–3
10. Fang, G., Zhang, Y.: 'Design of Dual-band Dual-polarized Spaceborne Precipitation Radar Antenna', Journal of Electronics & Information Technology. 2016, 38, (8), pp 1977–1983
11. Bringi, V.N., Chandrasekar, V.: 'Polarimetric Doppler weather radar: principles and applications' (Cambridge University Press, 2001, 636 pp.)
12. Pruppacher, H.R., Klett, J.D.: 'Microphysics of Clouds and Precipitation' (D. Reidel Pub. Co. 1978, pp. 381–382.)
13. Cao, J.W., Liu, L.P., Chen, X.H., et al.: 'Data Quality Analysis of 3836 C-Band Dual-linear Polarimetric Weather Radar and Its Observation of a Rainfall Process', Journal of Applied Meteorological Science, 2006, 17, (2), pp 192–200
14. Melnikov, V.M., Zrnic, D.S., Doviak, R.J., et al.: 'Prospects of the WSR-88D Radar for Cloud Studies', Journal of Applied Meteorology & Climatology, 2011, 50, (4)
15. Zhu, H.B., Pan, K.: 'Summery for the Structural Design of X Band Vehicle-borne Doppler Weather Radar System', Electro-mechanical Engineering, 2003, 19, (4), pp 8–10
16. Skolnik, M.I., Borkowski, M.T., Blake, L.V., et al.: 'Radar Handbook' (Radar Antennas, 2008.)
17. [17] Probert Jones, J.R.: 'The radar equation in meteorology', Quarterly Journal of the Royal Meteorological Society, 1964, 90, (383), pp 485–495
18. Zhang, P.C., Du, B.Y., Dai, T.P.: 'Radar Meteorology' (Beijing: China Meteorological Press, 2001.), pp. 14–15
19. Zhang, J.L., Wang, M., Tian, E.M., et al.: 'Analysis and experimental verification of sensitivity and SNR of laser warning receiver', Spectroscopy & Spectral Analysis, 2009, 29, (1), pp 20–23

A New 2D Doppler Steering Method Used for Format Flying Satellites InSAR System

Chipan Lai$^{(\boxtimes)}$, Aifang Liu, Dong Mu, Yixuan Zhao,
and Liang Cheng

No. 1 Research Department, Nanjing Research Institute of Electronics
Technology, Nanjing, China
wolf_lcp@sina.com

Abstract. Beam synchronization is a key technology for format flying satellites InSAR system, and the Doppler center frequency is a very important parameter for spaceborne SAR image formation. The more decrease of residual Doppler center frequency, the more convenience to SAR image formation. In this paper, the expression of Doppler center frequency is given after the analysis of the geometry of spaceborne SAR imaging. It considered the elliptical orbit and the pitch steering angle. For the sake of minimizing the Doppler center frequency, a new attitude steering method,so-called new zero Doppler steering, is suggested through the analysis of the expression of Doppler center frequency. The new method can minimize the residual Doppler centroid to absolutely zero Hz, and if all the satellites are in the formation applying this method, the beam synchronization of the system can be realized, and the common Doppler spectrum of tow images is maximal. The computer simulation results prove the validity of the method and show that the accuracy of new zero Doppler steering method is better than the total zero Doppler steering theoretically. The new zero Doppler steering method can be applied to the future spaceborne SAR and formation flying satellite InSAR system.

Keywords: Synthetic aperture radar · Format flying satellites · Beam synchronization · Attitude steering · New zero Doppler steerings

1 Introduction

A attitude steering method which is called total zero Doppler steering for spaceborne Synthetic Aperture Radar (SAR) system was described in paper [1–3]. The method can decrease the residual central Doppler frequency remarkably, and the steering law is same as right side-looking system and left side-looking system. This means less mutuality between azimuth signals and range signals. The less mutuality of azimuth and range signal, the more convenient for signal processing. Out of question, this is a quite meaningful method for spaceborne SAR system design. The method is applied first to German satellite SAR system TerraSAR-X, and the results accord with the expectation [4].

© Springer Nature Singapore Pte Ltd. 2019
L. Wang et al. (eds.), *Proceedings of the 5th China High Resolution Earth Observation Conference (CHREOC 2018)*, Lecture Notes in Electrical Engineering 552,
https://doi.org/10.1007/978-981-13-6553-9_10

In the paper, another attitude steering method is given. At first, the calculating expression of Doppler center frequency of spaceborne SAR is deduced through the analysis of the geometry of SAR imaging, which considered the elliptical orbit and the pitch steering angle. Second, through analysis of the expression, an improved attitude steering method—the so-called new zero Doppler steering—is suggested to minimize the residual Doppler centroid, as shown in Fig. 1; it can be used to realize the beam synchronization for format flying satellites InSAR system.

The new zero Doppler steering method compensates all error items of calculating expression of Doppler central frequency, but the total zero Doppler steering method is not, improving the performance of attitude steering. Without the attitude and other error sources, the residual Doppler center frequency is absolute zero Hz, theoretically. At last, the performance of different attitude steering methods is compared, including yaw steering, total zero Doppler steering, and new zero Doppler steering methods. The computer simulation results prove the validity of new zero Doppler steering method.

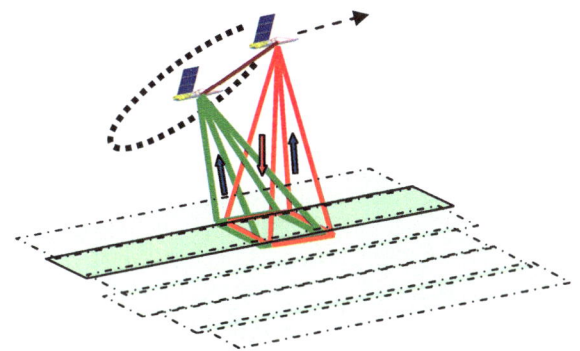

Fig. 1. Format flying satellites InSAR system and beam synchronization

2 The Expression of Doppler Central Frequency of Spaceborne SAR

The earth center coordinates system is shown in Fig. 2, where a—orbit semi-major axis; e—orbit eccentricity; i—orbit inclination angle; Ω—Right Ascension of Ascending Node (RAAN); ω—argument of perigee; f—true anomaly; α—argument of latitude, $\alpha = \omega + f$.

In the earth center coordinates system, the Doppler center frequency of spaceborne SAR f_{DC} can be expressed as follows [5–9]:

$$f_{DC} = -\frac{2}{\lambda R}[\mathbf{V_s} \cdot \mathbf{R} - \boldsymbol{\omega_e} \cdot (\mathbf{R_s} \times \mathbf{R})] \tag{1}$$

where λ—wavelength of center frequency; $\mathbf{R_t}$—earth radius vector of target; $\mathbf{V_s}$—satellite velocity vector; $\boldsymbol{\omega_e}$—earth rotation angle speed vector; $\mathbf{R}$—slant range vector; R—the absolute value of $\mathbf{R}$.

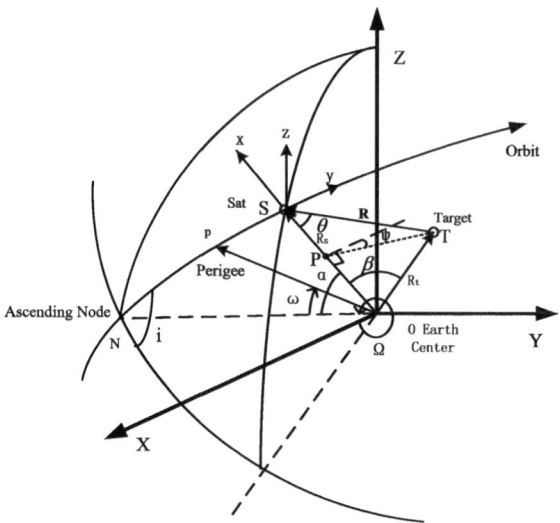

Fig. 2. Earth center coordinates system

If we get the expression of every vector in the satellite body coordinates system, the expression of Doppler center frequency will be acquired.

In the satellite body axis system $o - xyz$, denote its unit vector as $[\hat{\mathbf{u}}_\mathbf{r}, \hat{\mathbf{u}}_\mathbf{t}, \hat{\mathbf{u}}_\mathbf{p}]^\mathrm{T}$, then the expressions of those vectors are as follows [5, 6]:

$$\mathbf{R_s} = R_s \hat{\mathbf{u}}_\mathbf{r} \tag{2}$$

$$\mathbf{V_s} = \dot{R}_s \hat{\mathbf{u}}_\mathbf{r} + R_s \omega_s \hat{\mathbf{u}}_\mathbf{t} \tag{3}$$

$$\boldsymbol{\omega_e} = \omega_e \left(\sin i \sin \alpha \hat{\mathbf{u}}_\mathbf{r} + \sin i \cos \alpha \hat{\mathbf{u}}_\mathbf{t} + \cos i \hat{\mathbf{u}}_\mathbf{p} \right) \tag{4}$$

$$\mathbf{R} = R \left(\cos \theta_0 \hat{\mathbf{u}}_\mathbf{r} - \sin \theta_0 \cos \varphi_0 \hat{\mathbf{u}}_\mathbf{t} + \sin \theta_0 \sin \varphi_0 \hat{\mathbf{u}}_\mathbf{p} \right) \tag{5}$$

where R_s is the range from earth center to the satellite:

$$R_s = \frac{a \cdot (1 - e^2)}{(1 + e \cdot \cos f)} \tag{6}$$

$\dot{R}_s$ is the differential coefficient of R_s, $\mu = 3.986 \times 10^{14} \ \mathrm{m^3/s^2}$:

$$\dot{R}_s = \sqrt{\frac{\mu}{a(1 - e^2)}} \cdot e \cdot \sin f \tag{7}$$

ω_s is the satellite movement angle rate:

$$\omega_s = \sqrt{\frac{\mu \cdot a \cdot (1 - e^2)}{R_s^2}} \tag{8}$$

ω_e is the earth rotate angle rate, $\omega = 7.29211 \times 10^{-5}$ rad/s.

Now we can write out the expression of f_{DC} in an elliptical orbit. From (3), we can calculate the angle γ_0 between satellite velocity vector and the satellite body axis y, which is called track angle. We know that

$$\tan(\gamma_0) = \frac{\dot{R}_s}{R_s \omega_s} \tag{9}$$

Then, we get [7]:

$$\gamma_0 = \arctan\left(\frac{e \sin f}{1 + e \cos f}\right) \tag{10}$$

Here, we point out that in a circular orbit, the satellite velocity vector and the satellite body axis y are superposed, so γ_0 is zero. The relation of them is shown in Fig. 3.

For the purpose of deducing the new zero Doppler steering method, an axis transformation is performed. If we denote the pitch steering angle as γ, the axis after pitch steering as $o - x'y'z'$ and its unit vector as $\left[\hat{u}'_r, \hat{u}'_t, \hat{u}'_p\right]^T$, then we get the transform matrix from $o - xyz$ to $o - x'y'z'$, then we can write that

$$\begin{bmatrix} \hat{u}'_r \\ \hat{u}'_t \\ \hat{u}'_p \end{bmatrix} = \begin{bmatrix} \cos \gamma & \sin \gamma & 0 \\ -\sin \gamma & \cos \gamma & 0 \\ 0 & 0 & 1 \end{bmatrix} \cdot \begin{bmatrix} \hat{u}_r \\ \hat{u}_t \\ \hat{u}_p \end{bmatrix} \tag{11}$$

Make use of (11), (2)–(5) can be rewritten as

$$\mathbf{R_s} = R_s \cos \gamma \, \hat{u}'_r - R_s \sin \gamma \, \hat{u}'_t \tag{12}$$

$$\mathbf{V_s} = \left(\dot{R}_s \cos \gamma + R_s \sin \gamma\right) \hat{u}'_r + \left(-\dot{R}_s \sin \gamma + R_s \omega_s \cos \gamma\right) \hat{u}'_t \tag{13}$$

$$\omega_e = \omega_e \cdot (\sin i \sin(\alpha + \gamma)) \hat{u}'_r + \sin i \cos(\alpha + \gamma) \hat{u}'_t + \cos i \hat{u}'_p \tag{14}$$

$$\mathbf{R} = R \cdot \left(\cos \theta_0 \hat{u}'_r - \sin \theta_0 \cos \varphi_0 \hat{u}'_t + \sin \theta_0 \sin \varphi_0 \hat{u}'_p\right) \tag{15}$$

Use (1) and (12) to (15), in $o - x'y'z'$ coordinates system, the expression of Doppler centroid f_{DC} is as follows:

$$f_{DC} = -\frac{2}{\lambda} \left[\cos\theta_0 (\dot{R}_s \cos\gamma + R_s\omega_s \sin\gamma) \right.$$
$$- \sin\theta_0 \cos\varphi_0 (-\dot{R}_s \sin\gamma + R_s\omega_s \cos\gamma)$$
$$+ R_s\omega_e (\sin i \sin\theta_0 \sin\varphi_0 \cos\alpha$$
$$\left. + \cos i \sin\theta_0 \cos\varphi_0 \cos\gamma - \cos i \cos\theta_0 \sin\gamma) \right] \tag{16}$$

where: θ_0—look angle; φ_0—side-looking angle (range from 0–π). Equation (16) is the base of all attitude steering methods.

3   The Derive of New Zero Doppler Steering Method

To minimize the influence of Doppler centroid, we want to set f_{DC} to be zero Hz in (16). So we write that

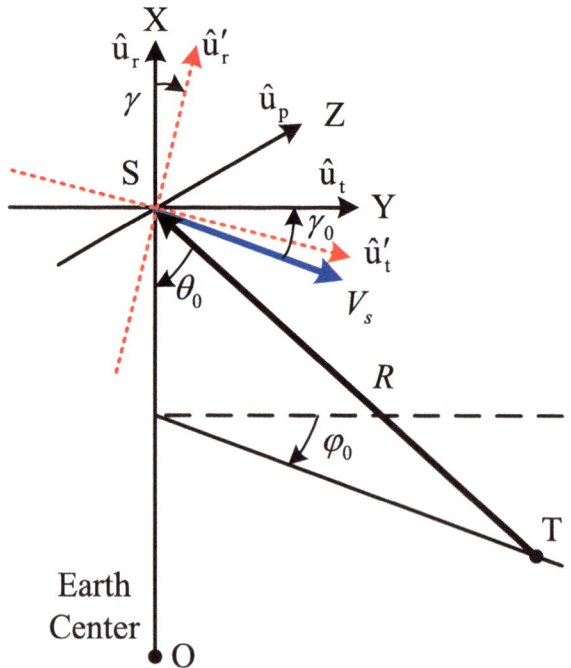

Fig. 3. Satellite body coordinates system

$$\dot{R}_s \cos\theta_0 \cos\gamma + R_s\omega_s \cos\theta_0 \sin\gamma - R_s\omega_e \cos\theta_0 \cos i \sin\gamma = 0 \tag{17}$$

$$R_s\omega_e \cos i \sin\theta_0 \cos\varphi_0 \cos\gamma + \dot{R}_s \sin\theta_0 \cos\varphi_0 \sin\gamma - R_s\omega_s \sin\theta_0 \cos\varphi_0 \cos\gamma$$
$$+ R_s\omega_e \sin i \sin\theta_0 \sin\varphi_0 \cos\alpha = 0 \tag{18}$$

Then, we can get

$$\gamma = -\arctan \frac{\dot{R}_s}{R_s(\omega_s - \omega_e \cos i)} \tag{19}$$

$$\phi = \arctan \frac{\sin i \cos \alpha}{\left(\frac{\omega_s}{\omega_e}\cos\gamma - \cos i \cos\gamma + \frac{\dot{R}_s}{R_s\cdot\omega_e}\sin\gamma\right)} \tag{20}$$

Equations (19) and (20) are the pitch steering angle and yaw steering angle of new zero Doppler steering method. Figure 4 shows the variety of them versus attitude of latitude.

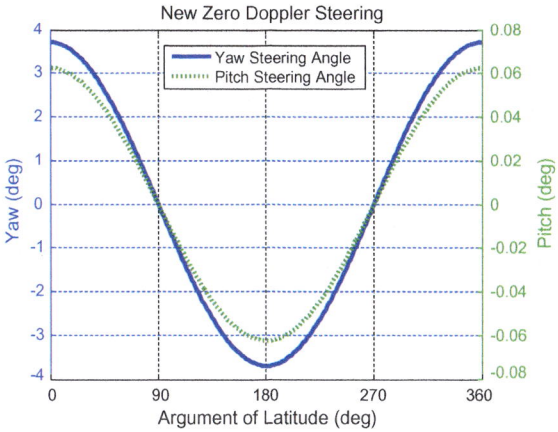

Fig. 4. New zero Doppler steering (NZDS) method law

4 Performance Analysis

Now we will show the performance of new zero Doppler steering method. Table 1 is the main system parameters of TanDEM-X satellite used in computer simulation [1–3].

Table 1. TanDEM-X system parameters

Parameter	Value
Semi-major axis	6892.137 km
Inclination	97.42°
Eccentricity	0.0011
Argument of perigee	90°
Look angle for yaw steering (mid-range)	33.8°
Look angle near range	18.45°
Look angle far range	49.25°
Repeat cycle N	167/11
Wavelength	0.03125 m

4.1   New Zero Doppler Steering Method

The new zero Doppler steering applies the pitch and yaw steering as (19) and (20), respectively. The residual Doppler center frequency is absolute zero Hz from the above analysis. The simulation results prove the validity of new zero Doppler steering method, as shown in Fig. 5.

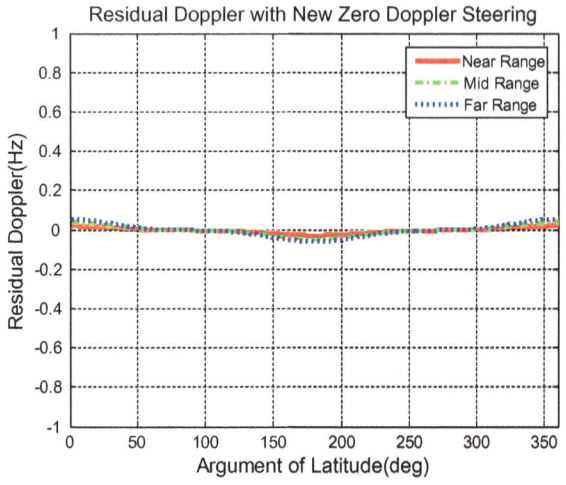

Fig. 5. Residual Doppler of new zero Doppler steering method

4.2   Total Zero Doppler Steering Method

The total zero Doppler steering method used (21) as the yaw steering angle, which is the approximation of (20), and an additional error is added to the residual Doppler center frequency because of the approximation. The residual Doppler center frequency of total zero Doppler steering method is shown in Fig. 6. The result is in a range of ±30 Hz.

$$\phi = \arctan \frac{\sin i \cos \alpha}{N - \cos i}, \quad N = \omega_s/\omega_e \tag{21}$$

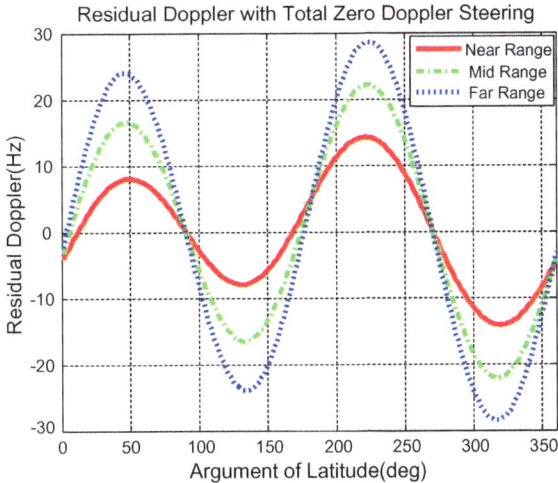

Fig. 6. Residual Doppler of total zero Doppler steering method

4.3 Performance Compare of Two Methods

Use the parameters in Table 1, and the simulation result is shown that the performance of new zero Doppler steering method is almost 0 Hz, but for the total zero Doppler steering method, the performance is ±30 Hz.

The new zero Doppler steering method is better than the total zero Doppler steering method, through our simulation results.

5 Conclusions

New zero Doppler steering is a new method applying to spaceborne SAR system to minimize the residual Doppler center frequency, and it can also be used to realize the beam synchronization of format flying satellites InSAR system. It is a new 2D attitude steering method, which enhances the performance of attitude steering. As the satellite attitude error has a big influence on the accuracy of attitude steering, higher accuracy of satellite attitude control will be requested. From the analysis and simulation results, we can make a conclusion that the performance of new zero Doppler steering method is better than the existing method, such as total zero Doppler steering. It has values both in theory and engineering in the area of spaceborne SAR. It can be used to the future spaceborne SAR and InSAR systems, especially, used to perform the beam synchronization of the formation flying satellite InSAR systems.

References

1. H. Fiedler, E. Boerner, J. Mittermayer, G. Krieger, "Total zero doppler steering", in Proceedings of the EUSAR 2004, Ulm, Germany, vol. 2, pp. 481–484, 2004.
2. H. Fiedler, E. Boerner, J. Mittermayer, G. Krieger, "Total zero doppler steering – a new method for minimising the doppler centroid", submitted to Geoscience and Remote Sensing Letters, 2004.
3. E. Boerner, H. Fiedler, G. Krieger, J. Mittermayer, A new method for total zero doppler steering, Geoscience and Remote Sensing Symposium, 2004. IGARSS '04. Proceedings. 2004 IEEE International.
4. Hauke Fiedler, Thomas Fritz; Ralph Kahle; "Verification of the total zero doppler steering", Radar, 2008 International Conference on, Sept. 2008.
5. J. C. Curlander, R. N. McDonough, Synthetic aperture radar: systems and signal processing, John Wiley & Sons, Inc., 1991.
6. Chipan Lai, Guangyan Liu, Youquan Lin, Qingfu Sun, "The effect of satellite attitude errors on spaceborne SAR doppler parameter", Modern Radar, 28 (9), pp. 15–17, 2006 [Modern Radar, In Chinese, 2006].
7. Zhongquan Wei, Synthetic Aperture Radar Satellite[M]. Beijing: The publishing company of science, 2001, pp. 132–152. [In Chinese].
8. Peng Zhou, Zi Xu, Yiming Pi, "Precise calculation of spaceborne SAR doppler parameter in ellipse orbit", Science of Mapping, Vol. 33, No. 6, 2008 [In Chinese].
9. R. K. Raney, "Doppler properties of radars in circular orbits," Int. J. Remote Sens., vol. 7, no. 9, pp. 1153–1162, 1986.

Research on Imaging Algorithms for Bistatic System of Geosynchronous SAR Based on Stationary Motion Mode

Dejin Tang[1]([⊠]), Xiaoming Zhou[2], Caiping Li[2], and Yuchen Song[2]

[1] National Geomatics Center of China, Beijing 100830, China
zxm2913@163.com
[2] Beijing Institute of Remote Sensing Information, Beijing 100192, China

Abstract. The geosynchronous orbit SAR system (GEOSAR) can monitor a certain area for a long time. It has the advantages of short revisit cycle and strong anti-attack ability. It has wide application prospect and development potential in military and civil fields. But at the same time, the time variability of the atmosphere in the long synthetic aperture affects the focusing performance of the L-band GEOSAR, and its antenna size, power consumption, and the cost of launching are also greatly limited to the engineering implementation of the radar. In order to solve the above problem, this paper derives the imaging algorithm of the double-base stationary motion model based on the concept of the binary radar system (GEO-UAV SAR) of the star machine hybrid system (GEO-UAV SAR) received by GEOSAR and carries out the simulation test. The basic theoretical model is put forward for the research of the imaging algorithm in the nonstationary state, and the development of the test star is made. And construction provides a basic theoretical research result.

Keywords: Remote sensing information system · GEOSAR imaging algorithm · Dechirp · Bistatic system

1 Introduction

Spaceborne synthetic aperture radar (Synthetic Aperture Radar, SAR) can be widely used in the fields of geological exploration, resource survey, disaster management, and military reconnaissance [1, 2]. At present, the spaceborne SAR on orbit is the low-orbit satellite (LEOSAR), which is within the range of tens or even hundreds of kilometers near the start point. Therefore, its coverage area is small, the surveying and mapping zone is narrow, and the repeated visit cycle is long, so it is difficult to meet the need of fine imaging in the local area. In order to solve this problem, improving the orbit height is an effective means to solve this restriction. Since the century, with the improvement of the antenna technology, load capacity, and computer performance, the feasibility of the spaceborne SAR from the low track to the high rail and even the synchronous orbit has been widely expected and paid more attention in the world. The geosynchronous orbit SAR (Geosynchronous Earth Orbit SAR, GEOSAR) imaging system has a wide swath band, short repeated access cycle, strong anti-attack, and anti-destruction ability, and can be monitored for a long time in specific areas relative to the low rail SAR [3, 4].

© Springer Nature Singapore Pte Ltd. 2019
L. Wang et al. (eds.), *Proceedings of the 5th China High Resolution Earth Observation Conference (CHREOC 2018)*, Lecture Notes in Electrical Engineering 552,
https://doi.org/10.1007/978-981-13-6553-9_11

Because of its large antenna size, high power consumption, and high emission cost, GEOSAR greatly restricts its engineering implementation. At present, there are still running GEOSAR satellites in the world. The signal processing and imaging algorithm model is still in the theoretical research stage. In order to solve the problem of the antenna and power bottleneck of the high-orbit SAR satellite, a dual-base structure (GEO-UAV SAR) satellite system based on GEOSAR as the launching platform and the unmanned airborne SAR (Unmanned Aerial Vehicle SAR, UAVSAR) is proposed. It can solve this problem well and can change the working mode of the receiving platform. The high resolution or wide swath imaging of different regions is realized. However, due to the elliptical motion characteristics of the launch platform GEOSAR, the geometric relationship of the hybrid radar is very complicated. In this paper, a GEOSAR imaging algorithm based on an ideal stationary operation model is explored, which provides an algorithm basis for the subsequent actual on-orbit nonstationary state imaging theory, so as to achieve the theoretical research aim of providing predictions for the test stars.

2 Bistatic System of GEOSAR Geometric Model

2.1 The Establishment of Local Ground Coordinates

The geometric configuration of the bistatic radar system is the basis of the echo signal model and imaging algorithm. Considering the speed of GEOSAR's motion is not equal to the speed of beam surface scanning, it is necessary to analyze the geometric configuration of GEO-UAV SAR in detail according to the beam surface scanning speed of the transceiver platform. Therefore, in this section, GEO-UAV SAR is described in the local ground coordinate system, and the geometric configuration of the dual-base radar is analyzed according to the ratio of the ground surface scan speed of the two platforms. This paper deduces the mathematical expression of GEOSAR's position and velocity in the local coordinate system.

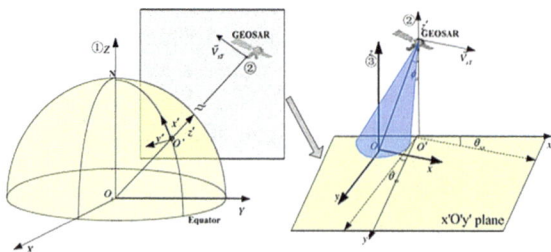

Fig. 1. Geometrical relationship between earth center fixed coordinate system and local ground coordinate system

In order to redescribe GEOSAR in the local terrestrial coordinate system, we must study the transformation relationship between the geocentric fixed coordinate system and the local terrestrial coordinate system, as shown in Fig. 1. The conversion process of coordinate system is geocentric fixed coordinate system, indirect local terrestrial coordinate system, and local terrestrial coordinate system. According to the Bohr Wolf model, the position coordinates of GEOSAR in the local terrestrial coordinate system are obtained.

$$F_{s\_L}(t_0 + t) = S_1[F_{O\_E}(t_0) - F_{O'\_E}(t_0)]^T + S_2 S_1[F_{s\_E}(t_0 + t_a) - F_{O'\_E}(t_0)] \quad (1)$$

$$S_1 = \begin{bmatrix} -\sin\Theta_{\text{lon}}(t_0)\cos\Theta_{\text{lat}}(t_0) & -\sin\Theta_{\text{lon}}(t_0)\sin\Theta_{\text{lat}}(t_0) & \cos\Theta_{\text{lon}}(t_0) \\ \sin\Theta_{\text{lat}}(t_0) & -\cos\Theta_{\text{lon}}(t_0) & 0 \\ \cos\Theta_{\text{lon}}(t_0)\cos\Theta_{\text{lat}}(t_0) & \cos\Theta_{\text{lon}}(t_0)\sin\Theta_{\text{lat}}(t_0) & \sin\Theta_{\text{lon}}(t_0) \end{bmatrix} \quad (2)$$

$$S_2 = \begin{bmatrix} \cos\theta_{sx} & \sin\theta_{sx} & 0 \\ -\sin\theta_{sx} & \cos\theta_{sx} & 0 \\ 0 & 0 & 1 \end{bmatrix} \quad (3)$$

In the local terrestrial coordinate system, the relative velocity between GEOSAR and the earth is

$$V_{s\_L}(t_0 + t_a) = S_2 S_1 V_{s\_E}(t_0 + t_a) \quad (4)$$

The surface scanning speed of the radar beam is as follows:

$$V_{gT}(t_0 + t_a) \quad (5)$$

2.2 Geometric Structure of Bistatic System

According to the beam scanning speed of the GEO-UAV SAR transceiver platform, the geometric characteristics of the dual-base radar are analyzed. It is assumed that the mathematical expressions of the velocity ratio of the beam ground Δ_1 scan to the velocity angle of the transceiver platform Δ_2 are

$$\begin{cases} \Delta_1 = (\varpi \cdot V_{gT}(t_0))/(\varpi \cdot V_R) \\ \Delta_2 = a\cos((V_{gt}(t_0) \cdot V_R)/|V_{gT}(t_0) \cdot V_R|) \end{cases} \quad (6)$$

The ϖ represents the access to unit vectors, $\varpi = \begin{bmatrix} 1 & 0 & 0 \end{bmatrix}^T$.

According to Δ_1 and Δ_2 to the numerical value, the geometric configurations of GEO-UAV SAR can be divided into five categories.

(1) $\Delta_1 = 0$ & $\Delta_2 = 0$, it indicates that the inclination angle of GEOSAR is zero and is at rest;
(2) $0 < \Delta_1 < 1$ & $\Delta_2 = 0$, it indicates that GEOSAR is in motion, the two platforms are parallel to the same route, and the GEOSAR beam's surface scanning speed is less than UAVSAR;

(3) $\Delta_1 \geq 1$ & $\Delta_2 = 0$, it indicates that GEOSAR is in motion, the two platforms are parallel to the same route, and the GEOSAR beam's surface scanning speed is greater than UAVSAR;

(4) $\Delta_1 < 0$ & $\Delta_2 = 0$, it is shown that GEOSAR is in motion state, the two platform parallel route reverse flights, and the geometric configuration change of the double-base radar caused by the mutual movement of the two platforms is similar to that of the third;

(5) $\Delta_2 \neq 0$, it indicates that GEOSAR is in motion and the two platforms are crossed. In this case, three geometric configurations can be further classified according to the Δ_1 numerical value. $\Delta_1 < 0$, $\Delta_1 > 1$, $0 < \Delta_1 < 1$, the surface beam motion is similar to the 2–4 case and is no longer duplicated.

3 Imaging Algorithm Based on GEO-UAV SAR

Under the condition of smooth motion of the GEO-UAV SAR transceiver platform, the two-dimensional spectrum of point target is derived based on the geometric configuration of the dual-base radar, and then the imaging algorithm is introduced.

3.1 Two-Dimensional Spectrum Derivation of Point Target

Figure 2 shows the geometric configuration of GEO-UAV SAR, and the expression of echo signal of GEO-UAV SAR for P_m point targets in the ground scene is as follows:

$$mm(t_r, t_a; p_m) = \rho_r\left(t_r - \frac{R(t_a; p_m)}{C}\right)\rho_a\left(t_a - \frac{X_m}{\mu V_R}\right)\exp\left(j\pi\gamma\left(t_r - \frac{R(t_a; p_m)}{C}\right)^2\right)$$
$$\exp\left(-j\frac{2\pi}{\lambda}R(t_a; p_m)\right) \tag{7}$$

The distance signal $ss(t_r, t_a; p_m)$ to the Fourier transform is applied to the echo signal, and the stationary-phase principle is applied to obtain the mathematical expression of the echo signal in the range frequency azimuth time domain.

$$mm(t_r, t_a; p_m) = \rho_r(f_r)\rho_a\left(t_a - \frac{X_m}{\mu V_R}\right)\exp\left(-j\pi\frac{f_r^2}{\gamma}\right)\exp\left(-j\frac{2\pi(f_c + f_r)}{C}R(t_a; p_m)\right) \tag{8}$$

Then, by means of azimuth Fourier transform, the MSR method is used to obtain the two-dimensional spectrum expression of the unified point target with space variation.

$$mm(t_r, t_a; p_m) = \rho_r(f_r)\rho_a(f_a)\exp(-j2\pi M_a(f_a; p_m))\exp(-j2\pi M_1(f_a; p_m)f_r)$$
$$\exp(-j2\pi M_2(f_a; p_m)f_r^2)\exp(-j2\pi M_3(f_a; p_m)f_r^3)\exp(-j2\pi M_s(p_m)) \tag{9}$$

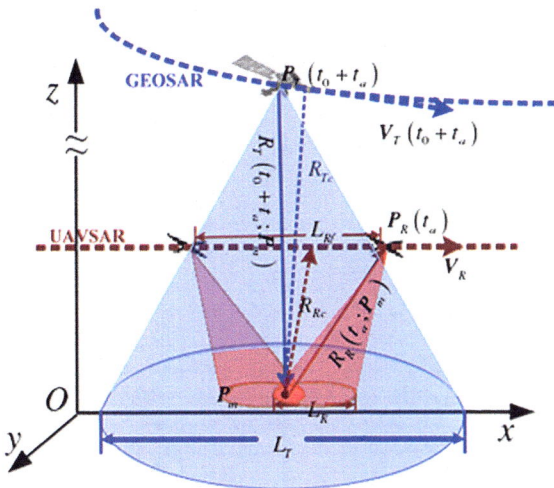

Fig. 2. Geometrical structure of GEO-UAV SA

3.2 Imaging Algorithm Flow

In this section, the Dechirp technology is introduced. Under the condition of not significantly increasing the amount of data, the unambiguous restoration of the scene when the PRF is slightly larger than the instantaneous Doppler bandwidth is completed, and the focus imaging of the whole scene is guaranteed. The imaging algorithm based on GEO-UAV SAR is shown in Fig. 3.

The algorithm process mainly contains the following steps:

(1) High-order phase compensation processing

From the two-dimensional spectrum expression of point target, there is a high-order distance and azimuth coupling function. In order to eliminate the image of the high-order coupling, the phase compensation function is multiplied in the two-dimensional frequency domain. The compensation function is

$$H_1(f_r, f_a; p_m) = \exp(j2\pi M_3(f_a; p_m)f_r^3) \tag{10}$$

(2) Azimuth-frequency domain aliasing processing

The dechirp technology is used to multiply the two-dimensional spectrum echo signal by the rotation phase function to eliminate the influence of antenna rotation on the Doppler frequency.

$$H_{\text{ref}}(t_a) = \exp(j\Omega\pi t_a^2), \ \Omega = \frac{(1-\mu)V_R^2 \cos^2\theta_{Rsq}}{\lambda R_{R0}} \tag{11}$$

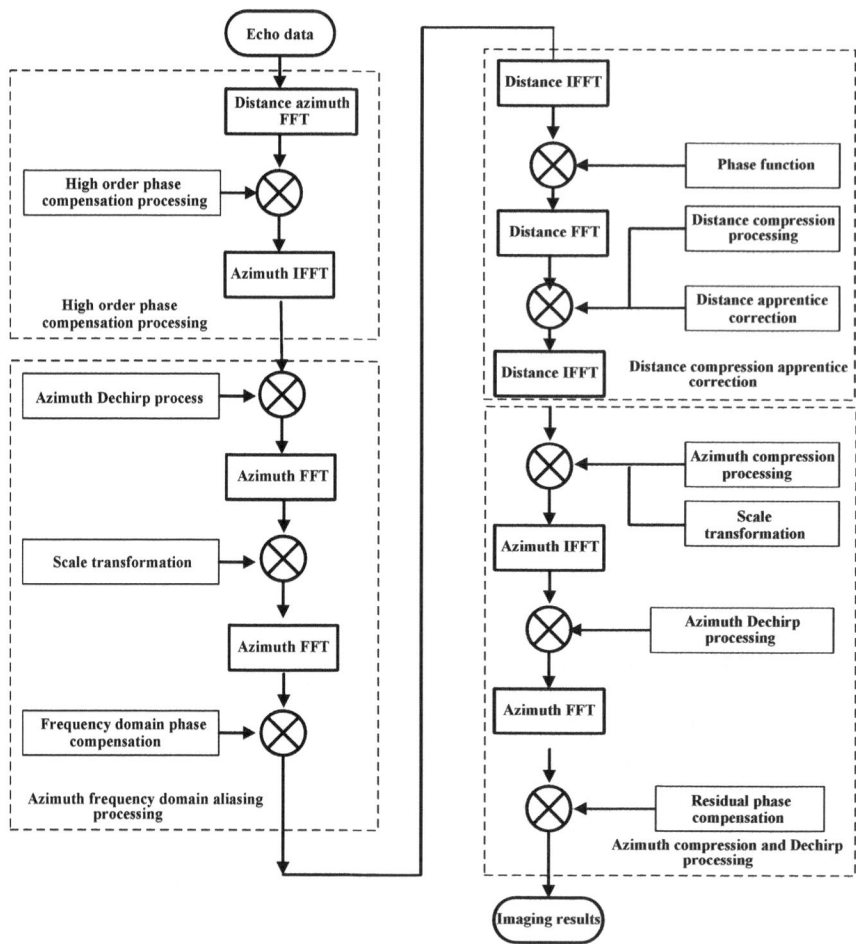

Fig. 3. Flowchart of imaging method for GEO-UAV SAR

(3) Distance compression and range migration correction

The two-dimensional frequency-domain signal $mm_2'(f_r, f_a'; p_m)$ is transformed by distance inverse Fourier transform, and the phase function is processed in the range-time azimuth-frequency domain, and the distance compression and distance migration correction are carried out in the two-dimensional frequency domain by the distance Fourier transform. The expression of the echo signal is as follows:

$$mm_3'(f_r, f_a'; p_m) = \sin c\left(B_r\left(t_r - \frac{R(t_0; p_m)}{C}\right)\right)\rho_a(f_a')\exp(-j2\pi M_a(f_a'; p_m))\exp(-j2\pi M_s(p_m))$$

$$(12)$$

(4) Azimuth compression and dechirp processing

The azimuth compression of the echo signal $mm'_3(f_r, f'_a; p_m)$ is carried out, and its compression function is

$$H_5(f'_a; p_m) = \exp\left(j2\pi\left(M_a(f'_a; p_m) - \frac{X_m}{V}f'_a\right)\right) \tag{13}$$

After processing by dechirp, the expression of echo signal is obtained.

$$mm'_5(t_r, f'_a; p_m) = \sin c\left(B_r\left(t_r - \frac{R(t_0; p_m)}{C}\right)\right)\sin c\left(\mu T_{\text{sar}}\left(f'_a + \frac{\Omega X_m}{\mu V_R}\right)\right) \\ \exp\left(j2\pi\frac{\Omega}{\mu}\left(\frac{X_m}{V_R}\right)^2\right)\exp(-j2\pi M_s(p_m)) \tag{14}$$

At this point, the echo signal p_m processing of the point target is completed.

4 Simulation Experiment

In order to demonstrate the effectiveness of this algorithm, nine point targets are set up in the simulation scene as shown in Fig. 4, and then the simulation and contrast test data are used for the open Radarsat data downloaded on the Internet (Table 1).

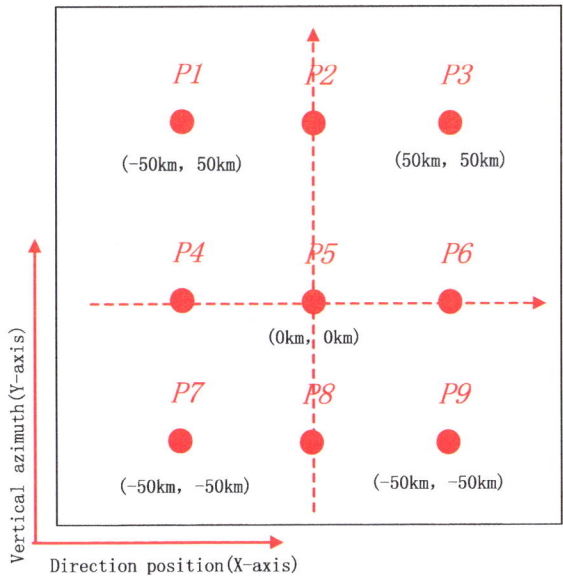

Fig. 4. Position of nine point targets in the simulate scene

Table 1. Imaging qualities of point targets for GEO-UAV SAR

UAVSAR mode	Target	Distance			Azimuth		
		PSLR (db)	ISLR (db)	IRW (m)	PSLR (db)	ISLR (db)	IRW (m)
Bunching	P1	−13.02	−10.11	2.01	−13.16	−10.18	0.17
	P5	−13.21	−10.14	2.01	−13.19	−10.23	0.14
	P9	−13.03	−9.97	2.01	−13.13	−10.20	0.14
Slipping bunching	P1	−12.92	−10.24	2.01	−13.27	−10.17	0.24
	P5	−13.32	−10.27	2.01	−13.15	−10.22	0.26
	P9	−13.23	−10.18	2.01	−12.99	−10.18	0.27
TOPS	P1	−13.23	−10.10	2.01	−13.20	−10.05	2.01
	P5	−13.22	−10.14	2.01	−13.22	−10.14	2.01
	P9	−13.23	−10.07	2.01	−13.18	−10.07	2.11

Subsequently, the hardware in the loop simulation experiment is carried out using the Radarsat raw data downloaded from the Internet. Figure 5a gives the focusing results of the Radarsat raw data. Figure 5b shows that this image is consistent with the original SAR image resolution by imaging simulation results obtained in this paper, which proves that the proposed imaging algorithm in this section can deal with multipoint target echo better.

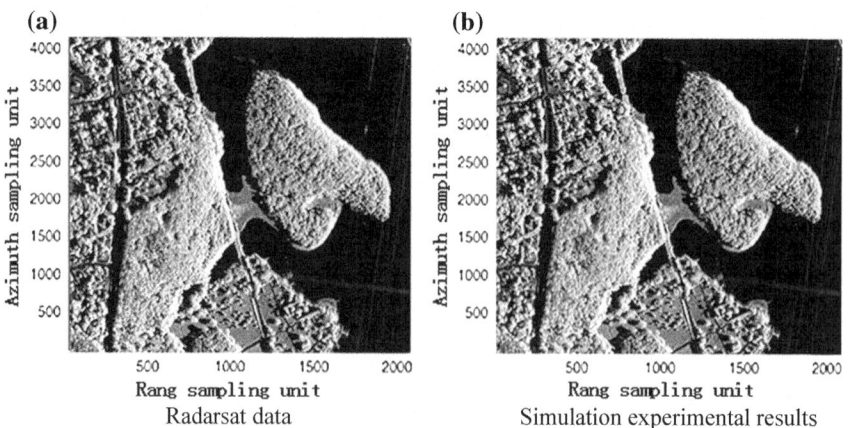

Fig. 5. Imaging result of semi-physical simulation

5   Conclusion

In this paper, the imaging algorithm of GEO-UAV SAR is mainly studied. First, based on the geometric structure relationship of GEO-UAV SAR, the MSR method is applied to deduce the unified two-dimensional spectral expression of point target. In view of the azimuth spectrum aliasing caused by the rotation of the receiving platform antenna, the dechirp technology is introduced and the imaging algorithm flow is proposed. The simulation results show that the method can better solve the aliasing problem caused by antenna rotation and obtain the focused imaging results.

References

1. Tomiyasu K, Pacelli J L. Synthetic Aperture Radar Imaging from an Inclined Geosynchronous Orbit [J]. IEEE Transactions on Geoscience and Remote Sensing, 1983, 21(3): 324–329.
2. Bruno D, Hobbs S E, Ottavianelli G. Geosynchronous Synthetic Aperture Radar: Concept Design, Properties and Possible Applications [J]. Acta Astronautica, 2006, 59:149–156.
3. Rodon J R, Broquetas A, Guarnieri A M, et al. Geosynchronous SAR Focusing with Atmospheric Phase Screen Retrieval and Compensation [J]. IEEE Transactions on Geoscience and Remote Sensing, 2013, 51(8): 4397–4404.
4. Kou L, Wang X, Xiang M, Zhu M. Interferometric Estimation of Three-dimensional Surface Deformation Using Geosynchronous Circular SAR [J]. IEEE Transactions on Aerospace and Electronic Systems, 2012, 48(2):1619–1635.

Ultrashort-Term Forecasting of Mid–Low-Altitude Wind Based on LSTM

Chunli Chen[1], Jie Zhou[1], Xiaofeng Li[1], Peie Zhao[1], Tao Peng[1],
Guohua Jin[1], Zhiping Liu[1], Yan Zhang[2(✉)], Yong Chen[1],
and Xiong Luo[1]

[1] Southwest Institute of Technical Physics, Chengdu 610041, Sichuan, China
[2] School of Electronic and Communication Engineering, Guiyang University,
Guiyang 550005, Guizhou, China
Eileen_zy001@sohu.com

Abstract. Accurate ultrashort-term forecasting of mid–low-altitude wind is essential to the safe and stable flight of aircraft. Traditional artificial intelligence (AI) wind forecasting methods convert the dynamic time series regression problem into a static spatial modeling problem, while ignoring the dynamic characteristics of the wind as a typical time series, and the prediction accuracy is limited. In this paper, the long short-term memory (LSTM) network is used to dynamically model the time series of wind speed to realize ultrashort-term forecasting at mid–low altitudes. The measured data of wind lidar are used to verify the conclusions. The results show that the performance of LSTM model outperforms artificial neural network and support vector machine.

Keywords: Wind forecasting · Long short-term memory · Artificial intelligence · Wind lidar

1 Introduction

Wind is one of the important factors which affect the position and track of vehicle [1], and both theory research and flight test show that the change of the velocity vector is one of the important factors that cause flight path deviation. Due to the fluctuating and intermittent characteristics of the wind, the wind field is often complicated, and there are many kinds of changes such as constant wind, shear wind, and random wind. It is an important flight control disturbance source for the aircraft, especially the unmanned aerial vehicle (UAV) flying in the middle and low altitude and relatively low flying speed, which directly affects the safety of the aircraft [2, 3]. The short-term prediction of low and medium wind fields is of great significance for improving the safety of aircraft.

Foundation items: National Natural Science Foundation of China (61505036), the Fund Project of Guizhou Provincial Science and Technology Department (QKHJZ [2015] 2009).

L. Wang et al. (eds.), *Proceedings of the 5th China High Resolution Earth Observation Conference (CHREOC 2018)*, Lecture Notes in Electrical Engineering 552,
https://doi.org/10.1007/978-981-13-6553-9_12

At present, there are mainly methods [4, 5] for short-term prediction of wind field: statistical method, physical method, and learning method. The statistical method cannot accurately measure the time-periodic characteristics of wind field and the airspace distribution characteristics, and the prediction error is large. The physical method is based on the physical model of the wind field. In the actual modeling process, the prerequisites are often assumed, and there is a strong regional limit. It is not suitable for the real-time prediction of the flight area of the aircraft. Learning method is generally based on the machine learning method of artificial intelligence, learning and predicting the numerical characteristics of the training data directly, effectively avoiding the establishment and verification of the physical model. It is the most effective method of wind field prediction at present. The traditional artificial intelligence wind field prediction methods (such as RNN, SVM, etc.) often transform the dynamic time series regression to the static space modeling problem, ignoring the dynamic characteristics of the wind field as a typical time series, and the prediction accuracy is limited. In order to improve the prediction accuracy, this paper explores the dynamic time modeling of the nonlinear relationship between time series and wind speed sequence using LSTM network and constructs a prediction model. The backpropagation through time (BPTT) is used to effectively train the network parameters, and finally the ultrashort-term prediction of wind field is completed.

The results show that the method used in this paper can make use of time series information to predict wind field effectively. The prediction accuracy is better than that of RNN and support vector machine, and the prediction accuracy is improved.

2   Theory

Long short-term network is generally called LSTM, which is a special type of recurrent neural network (RNN), RNN is a general name for a series of neural networks that can handle sequence data, and LSTM can learn long-term dependence information. LSTM was proposed by Hochreiter and Schmidhuber [6] and was modified and promoted by Graves [7]. In many problems, LSTM has achieved considerable success and has been widely used. LSTM is designed to avoid long-term dependence. Remember that long-term information is the default behavior of LSTM in practice rather than the ability to acquire it at great cost.

Usually, the RNN contains the following three characteristics:

Recurrent neural network can generate one output at each time node, and the connection between hidden units is cyclic.

Recurrent neural network can produce an output at each time node, and the output of the time node is only circularly connected to the hidden unit of the next time node.

Recurrent neural network contains hidden units with cyclic connections and can process sequential data and output single forecast.

Because RNN calculates the connections between distant nodes, multiple multiplication of the Jacobi matrix leads to the problem of gradient disappearance (often occurring) or gradient expansion (less occurring). In order to solve this problem, researchers put forward many solutions. The most successful and widely used is the threshold RNN (Gated RNN), and LSTM is the most famous one in threshold RNN. By designing the weight coefficient between the connections, the leaky units allow the RNN to accumulate the long-term connections between distant nodes, allow the coefficient to be changed at different times, and allow the network to forget the current accumulated information.

LSTM is composed of input layer, hidden layer, and output layer (Fig. 1). Among them, $g(t)$ represents the input unit, $h(t)$ represents the state output unit, M represents memory unit, and $i(t)$, $o(t)$, $f(t)$ represent input threshold, output threshold, and forgetting threshold [8].

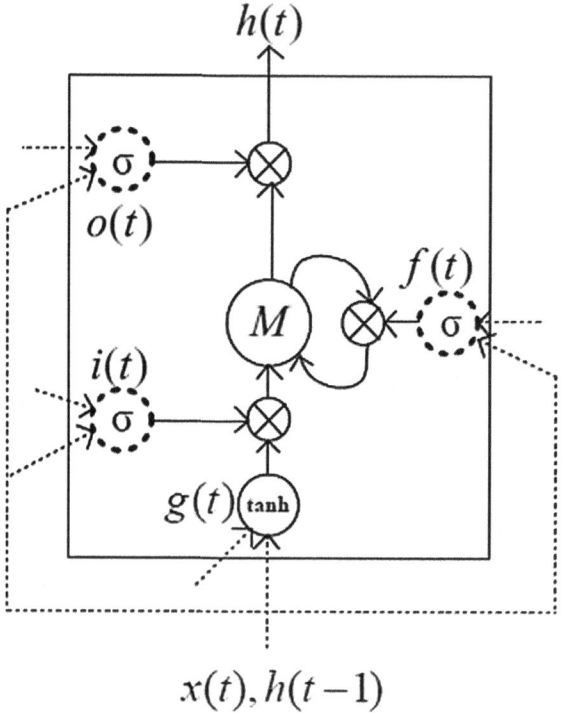

Fig. 1. LSTM structure

As can be seen from Fig. 1, the read, write, and forget operations of M are controlled by three threshold units. The input time sequence is x, t is the current time, and sigma represents the sigmoid activation function, and the state of each unit can be represented by the following formula.

Input unit:

$$g(t) = \tanh\left(W_{xg}g(t-1) + W_{kg}h(t-1) + b_g\right) \tag{1}$$

Gate control unit:

$$i(t) = \sigma(W_{xi}i(t-1) + W_{hi}h(t-1) + b_i) \tag{2}$$

$$f(t) = \sigma(W_{xf}i(t-1) + W_{hf}h(t-1) + b_f) \tag{3}$$

$$o(t) = \sigma(W_{xo}i(t-1) + W_{ho}h(t-1) + b_o) \tag{4}$$

Memory unit:

$$M(t) = f(t)M(t-1) + i(t)g(t) \tag{5}$$

State output unit:

$$h(t) = o(t)\tanh(M(t)) \tag{6}$$

It can be seen that the characteristic of LSTM is that by increasing input threshold, forgetting threshold and output threshold, the weight of self-circulation is changed, and the integral scale of different times can be changed dynamically when the model parameters are fixed, thus the problem of gradient disappearance or gradient expansion is avoided.

3   Prediction Model

3.1   Training Algorithm for LSTM Network

There are usually two methods for LSTM and other recurrent neural network models: one is backpropagation through time (BPTT) [9] and the other is real-time recurrent learning (RTRL) [10]. Because of its clear concept and efficient in computation, BPTT algorithm has more advantages than RTRL in computation time. Therefore, BPTT is used to train LSTM network in our paper.

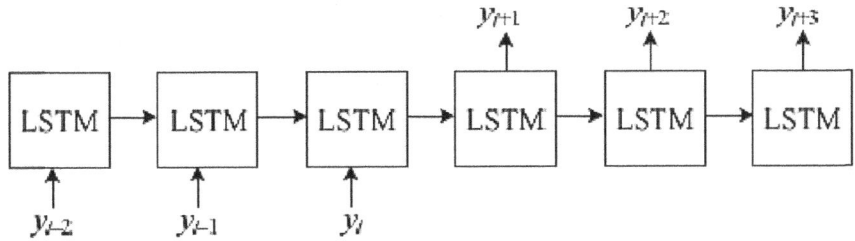

Fig. 2. LSTM training by BPTT

The basic idea of the BPTT algorithm is that the LSTM network is first expanded in a time sequence into a deep network, and then the classical backpropagation (BP) algorithm is used to train the expanded network. The intention is shown in Fig. 2. Like the standard BP algorithm, BPTT also needs to apply chain rules repeatedly. It should be noted that for LSTM network, the loss function is related not only to the output layer but also to the hidden layer of the time point.

3.2   Data for Prediction Model

The prediction model uses historical wind speed data as input to predict future wind speed. The data input and output structures of the model are shown in Table 1. X_i indicates the wind speed at the same height level at time i.

Table 1. Model input and output data structure table

Input data	Output data
$\{X_1, X_2, X_3, X_4, \cdots, X_n\}$	$\{X_{n+1}\}$
$\{X_2, X_3, X_4, X_5, \cdots, X_{n+1}\}$	$\{X_{n+2}\}$
$\vdots$	$\vdots$
$\{X_m, X_{m+1}, X_{m+2}, X_{m+3}, \cdots, X_{m+n}\}$	$\{X_{m+n+1}\}$

Among them, the dimension n of input data is adjustable parameter which directly affects the effect of prediction model. In this paper, the historical data of the wind lidar are used as the data sample. The selected wind lidar can measure the wind in multiple layers of the middle and low altitude, and a set of wind speed data is measured at each 2S. It is concluded from experiment that when n is selected in the range 8–12, the prediction model has a good effect. In this paper, $n = 10$ is taken to carry out the prediction test.

3.3   Evaluation Index

Four different evaluation indexes are adopted for prediction results in this paper. They are mean absolute error (MAE), root mean square error (RMSE), Nash–Sutcliffe efficiency coefficient (NSE), and statistical prediction accuracy (Spa).

$$\text{MAE} = \frac{1}{N}\sum_{i=1}^{N}|X_i - \hat{X}_i| \tag{7}$$

$$\text{RMSE} = \sqrt{\frac{1}{N}\sum_{i=1}^{N}(X_i - \hat{X}_i)^2} \tag{8}$$

$$\text{NSE} = 1 - \frac{\sum_{i=1}^{N}(X_i - \hat{X}_i)^2}{\sum_{i=1}^{N}(X_i - \bar{X})^2} \tag{9}$$

$$\text{Spa} = \frac{1}{N}\sum_{i=1}^{N}f(i) \tag{10}$$

$$f(i) = \begin{cases} 1, \left|X_i - \hat{X}_i\right|/X_i < 0.05 \\ 0, \left|X_i - \hat{X}_i\right|/X_i \geq 0.05 \end{cases} \tag{11}$$

X_i is the true data at time i, $\hat{X}_i$ is the prediction result of X_i, and $\bar{X}$ is the mean of X_i all over the time. The range of NSE is minus infinity to 1. When NSE is closer to 1, the quality of model is better and the credibility of model is higher. When NSE is close to 0, the simulation result is close to the average level of the observed value, that is, the overall result is believable, but the process simulation error is large. When NSE is far less than 0, the model type is unbelievable.

4   Results

The experimental data were measured by a wind lidar on the roof of a building in Chengdu urban area began at 11 a.m. on May 20, 2017. The test data were randomly selected from three high levels of 550, 850, and 1200 m, which started from 11:14:02. Among them, the measurement data of the first 5000 measurement time are used for the algorithm training according to 3.2, and the latter 500 are used for the result prediction. The prediction results of the three algorithms are shown in Fig. 3.

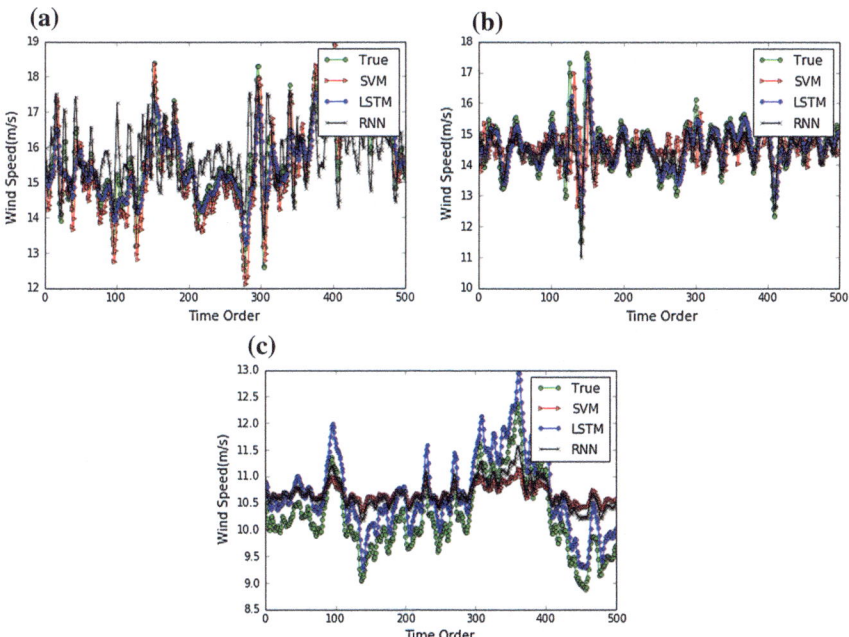

Fig. 3.   Prediction results. **a** 500 m; **b** 850 m; **c** 1200 m

The evaluation indexes of each algorithm are shown in Tables 2, 3, and 4. The prediction results of the LSTM model are better than the other models in all high level. The prediction results of the two models RNN and SVM are not stable, and there are great differences in the prediction results at 500 and 850 m levels. It can be seen that LSTM model has obvious advantages over RNN and SVM. In particular, when the wind speed steady changes in one direction at a period of time (such as 300–500 in Fig. 3c), the results of RNN and SVM prediction have obvious errors. While LSTM network can avoid the disappearance or expansion of the gradient in the transmission and span, and can also establish the interdependence of the time series of the wind, which leads to better prediction in this case.

Table 2. Evaluation indexes of each algorithm at 500 m

	MAE	RMSE	NSE	Spa
LSTM	0.404	0.551	0.780	0.868
RNN	0.666	0.786	0.554	0.61
SVM	0.451	0.584	0.753	0.854

Table 3. Evaluation indexes of each algorithm at 850 m

	MAE	RMSE	NSE	Spa
LSTM	0.300	0.419	0.721	0.924
RNN	0.351	0.467	0.652	0.904
SVM	0.582	0.803	0.271	0.708

Table 4. Evaluation indexes of each algorithm at 1200 m

	MAE	RMSE	NSE	Spa
LSTM	0.434	0.466	0.546	0.69
RNN	0.528	0.611	0.219	0.48
SVM	0.609	0.699	0.203	0.424

5   Conclusion and Further Work

In this paper, LSTM model is used to predict the wind of mid–low altitude in ultrashort-term. The experiment proves that the LSTM model has high prediction accuracy and good stability than RNN and SVM model. It provides a new idea for the ultrashort-term prediction of mid–low-altitude wind. There are many adjustable parameters in the LSTM model; improving the prediction accuracy by optimizing the adjustable parameters is our next research goal. In addition, there is still space to improve the prediction accuracy of machine learning prediction methods based solely on historical data. Combined with other meteorological information to correct the prediction results will be helpful for further improvement of prediction accuracy.

References

1. QU Yanlu. Ballistae Meteorology [M]. Beijing: China Meteorology Press, 1987, 162–165.
2. Lei Xusheng, Tao Ye. Adaptive Control for Small Unmanned Aerial Vehicle Under Wind Disturbance [J]. Acta Aeronautica Et Astronautica Sinica, 2010, 31(6): 1171–1176.
3. He YL, Chen YM, Zhou M. Modeling and control of a quadrotor helicopter under impact of wind disturbance[J]. Journal of Chinese Inertial Technology, 2013, 21(5): 624–630.
4. Costa A, Crespo A, Navarro J, et al. A review on the young history of the wind power short-term prediction [J]. Renewable and Sustainable Energy Reviews, 2008, 12(6): 1725–1744.
5. Yang Xiuyuan, Xiao Yang, Chen Shuyong. Wind speed and generated power forecasting in wind farm[J]. Proceedings of the CSEE, 2005, 25(11): 1–5.
6. Hochreiter S, Schmidhuber J. Long short-term memory [J]. Neural computation, 1997, 9(8): 1735–1780.
7. Graves A, Wayne G, Danihelka I. Neural turing machines [J]. arXiv preprint arXiv:1410. 5401, 2014.
8. Zhang Peng, Yang Tao, Liu Yananet al. Feature extraction and prediction of QAR data based on CNN-LSTM [J/OL]. Application Research of Computers, 2019(10):1–7.
9. Bengio Y, Simard P, Frasconi P. Learning long-term dependencies with gradient descent is difficult [J].IEEE Transactions on Neural Networks, 1994, 5(2):157–166.
10. Chang F, Chang C, Huang H, et al. Real-time recurrent learning network for stream-flow forecasting [J]. Hydrological Processes 2002,16(13):2577–2588.

Research on Formal Modeling and Analysis of Complex Space Information Systems

Huange Yin[1,2(✉)], Yanfeng Hu[2], and Liang Liu[2]

[1] School of Computer Science and Software Engineering, East China Normal University, Shanghai 200062, China
51164500262@stu.ecnu.edu.cn
[2] Institute of Electronics, Chinese Academy of Sciences, Suzhou 215123, Jiangsu, China

Abstract. As the scale and complexity of space information systems are continuously expanding, it is necessary to guarantee the system security and control the development cost as early as possible. According to the spatiotemporal characteristics of complex space information systems, this paper has proposed a formal method for modeling and analyzing the requirements of space information systems. First, a Space Information System Description Language, shorted as SISDL, is designed to formally describe system requirements. Then the corresponding tool for modeling system requirements by analyzing specific requirement document is developed. The well-built model is graphically displayed and used to check the existing errors in the requirement document. Finally, a ground information system of a specific satellite has been taken as an example for verification. The experimental results show that the method could effectively model the system requirements, and could improve the quality of the requirement document by analyzing the ambiguity in the requirement document.

Keywords: Space information systems · Requirement analysis · Formal method · Graphical · Modularization

1 Introduction

With the development of spatial information technology, the scale and complexity of spatial information system keep on increasing. The cost of manpower and material resources during the software development process advances with the times, and the security of its system is of great essence. At present, the commonly used method to ensure the security is to test the reliability and consistency of the system through the system test [1]. However, it is more expensive to modify the system when errors are found in the later test stage [2]. The cost can be better controlled if design errors can be found in the earlier requirement analysis stage [3–5]. The Formal Modeling Method [6] can be used for analysis in the requirement stage, but the existing Z [7], B [8], Event-B [9] and other methods are more suitable for embedded systems. Moreover, the learning cost is too high to be suitable for developers. The complex space information system is

L. Wang et al. (eds.), *Proceedings of the 5th China High Resolution Earth Observation Conference (CHREOC 2018)*, Lecture Notes in Electrical Engineering 552,
https://doi.org/10.1007/978-981-13-6553-9_13

a process-oriented modular system. In the stage of requirement analysis, engineers prefer to understand the relationship between modules of the system and consider the information flow and interface relationship between different modules. However, the requirement analysis described by natural language cannot ensure that it is consistent with the expectations of engineers. In this paper, a requirement description language shorted as SISDL is proposed to meet this requirement, and corresponding tools are designed to help developers model the system and check whether the design scheme is correct or not.

2   Requirement Modeling and Analysis Framework

The formal method is a modeling method based on mathematical conventions to reduce ambiguity [10], which can be used for the analysis of software requirement models. However, the high learning cost limits its application in industry. The SOLF [11, 12] method designed by Shaoying Liu has proposed a nonformal, semiformal, and formal progressive modeling concept that can reduce the difficulty for developers to learn the formal method. Based on this concept, the general framework of formal requirement modeling and analysis for space information systems is proposed (Fig. 1).

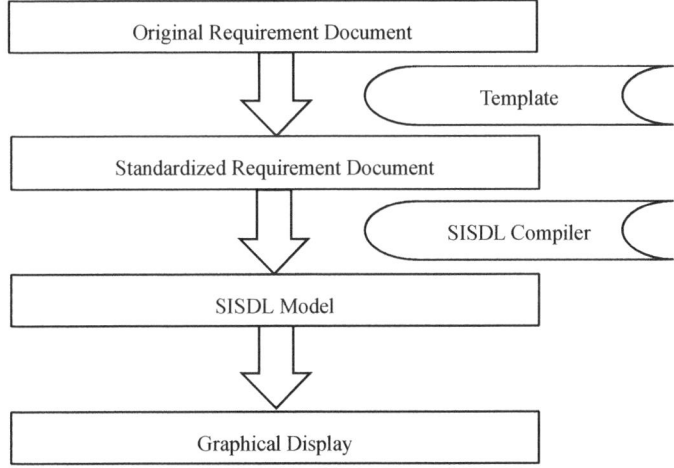

Fig. 1. General framework

The informal requirement document described by traditional natural language is not suitable for analysis. In this paper, a template for requirement document is designed to assist developers in compiling requirement document and obtain standardized requirement document. The standardized requirement document is read by the tool and converted into SISDL code format. The compiler analyzes the code to obtain SISDL model. The graphical display of the model shows the relationship between system modes and tasks. It is convenient for engineers to understand the system and check whether the design scheme is correct or not. If the system structure displayed graphically is different from the original intention, it is necessary to modify the requirement document to perfect the design scheme.

3   Space Information System Description Language

The complex space information system is a process-oriented and modular step-by-step system. SISDL is designed to describe the characteristics of decreasing granularity. It focuses on the interface relationship between modules. In this paper, SISDL is introduced from the perspectives of model structure and grammar design.

(1) **Model Structure**

According to the characteristics of the complex space information system, this paper divides the system model into two main levels. The mode is used to represent the subsystems of larger granularity, while the task is used to represent the functional modules of smaller granularity.

Mode structure
Mode:{ID, init, task}

where, "ID" refers to the name of the mode; "init" refers to the initialization of the mode, i.e., the initialization of the variables (the interface between the modules); "task" refers to the task of the mode call and the function modules contained in the subsystem.

Task structure
Task: {ID, time, space, input, proc, task, output}

where, "ID" refers to the name of the task; "time" refers to the time attribute of the task; "space" refers to the space attribute of the task; "input" refers to the input of the task; "proc" refers to the formula of the task, i.e., the specific operation of the variables; "task" is the task called; "output" is the output of the task.

The complex space information system has spatio-temporal characteristics. That is to say, the system's products and other operating objects have temporal and spatial attributes, such as (product_1, 2017-10-10, Beijing), the first-level image products generated for the Beijing area on 2017-10-10. The production module produces the product and the operation object serves as the product variable. The spatial parameters of the product are obtained from "space", the temporal parameters of the product are obtained from the current time of the system, and the product_ID, product_time, and product_space are uniformly saved. When the query module queries products, it searches the product through ID tag matching, judges whether the time and space of the ID meet the query requirements, and then returns the query results.

(2) **Grammar Design**

Considering the gradual refinement of the original requirement document, both modes and tasks are treated as functions in SISDL. For example, the subsystem is a function that includes initialization. The module is a function that has specific operations. It is convenient for engineers to convert the original requirement document. This paper gives Antlr [13] format of SISDL grammar.

This paper starts with the code block and uses the automatic downward approach to design the grammar.

Code block grammar
block
: (statement | functionDecl)* (Return expression ';')?
;

The **block** is the main body of the language and can be a statement or a function declaration. It can be ended with a return statement.

The **statement** contains declaration statement, function call statement, if statement, for statement, and while statement. Declaration statements and function calls end with a semicolon.

Statement grammar

statement
: assignment ';'
| functionCall ';'
| ifStatement
| forStatement
| whileStatement
;

Assignment is used for variable initialization of the mode

assignment
: Id indexes? '=' expression
;

FunctionCall calls the specified function via **Id**, which is used for mode and task calls

functionCall
: Id '(' exprList? ')'
;

IfStatement, **forStatement**, and **whileStatement** are used for formula description of the task.

If statement grammar

ifStatement
: ifStat elseIfStat* elseStat? End
;
ifStat
: If expression Do block
;
elseIfStat
: Else If expression Do block

;
elseStat
: Else Do block
;
For statement grammar:
forStatement
: For Id '=' expression To expression Do block End
;

The loop condition in the **forStatement** is only available for integers, such as "for a = 1–10". The first expression is the variable initialization and the second is the termination condition.

While statement grammar

whileStatement
: While expression Do block End
;

Function declaration grammar
functionDecl
: Def Id '(' idList? ')' block End
;

According to the model structure and grammar of SISDL, engineers can describe the design scheme of complex space information system according to the original requirement document, retain its characteristics of gradual refinement and focus on the interface relationship between modules.

4   Standardized Requirement Document

A Template for Requirement is designed to reduce the difficulty of engineers in writing standardized requirement documents. Developers convert the original requirement document into standardized requirement document according to the template. The template consists of three parts: variable dictionary, mode, and task. The main concern in the document is the interface relationship between different modules. In this paper, the interface in the document is defined as a variable. The name, type, and type description of all variables are given in the variable dictionary. It is convenient for developers to refer to when writing standardized requirement document (as shown in Table 1).

Table 1. Variable Dictionary

Variable Name	Variable Number	Variable Type	Type Description	Initial Value
String	ID Identifier	Type	String	...
...	...	...	...	...

The template contains models and tasks. The template follows the hierarchical characteristics of document writing. The subsystems are modes. The functional modules of the subsystem are tasks. It is convenient for developers to write documents more intuitively (Fig. 2).

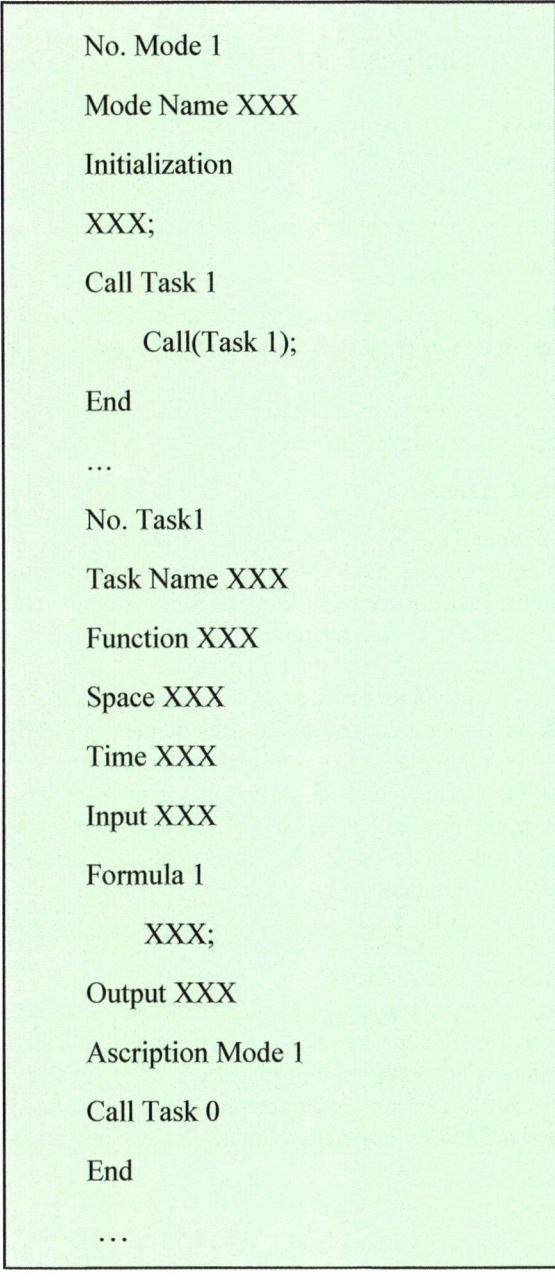

Fig. 2. Template for requirement document

According to the format of the template and the grammar specification of SISDL, the original requirement document is converted into a standardized requirement document. The tool processes the document to obtain the corresponding SISDL code. The code is analyzed and the system structure information are obtained through SISDL compiler.

5 Formal Visual Analysis

The internal structure of the system is shown through the mode call diagram and task call diagram to assist engineers to view the system architecture and internal operation process. Through processing the standardized requirement document, the relationships between modes and tasks are obtained and displayed graphically.

(1) **Mode call relationship**

A list of **callTasks** is defined. All tasks of the current mode are acquired by the task call of mode A. The ascription of the task is viewed. When the task does not belong to the current mode but to the mode B, it indicates that there is an association between mode A and mode B. All the calling relationships of the mode in call tasks are saved in **callTasks** for graph generation.

(2) **Task call relationship**

A list of **taskFunctions** is defined. The formula of the task has calls to other tasks, which indicates that there is an association between the two. All the calling relationships are saved in the **taskFunctions** of the task for graph generation.

The mode call shows the calling relationship between the subsystems in the system. When there is an interaction between the two subsystems, i.e., one subsystem has the task of calling another subsystem, the structure of the system at the subsystem level can be viewed via the relationship between these subsystems can clearly view.

The task call shows the relationship between the functional modules in the system. It can display specific information about the calling relationship between modes. In the task call, the relationship between specific functional modules between subsystems can be observed. Hence, it is convenient for engineers to check whether there are incorrect calls between functional modules.

6 Verification via Case Study

In order to carry out formal modeling, analysis and verification, the ground information processing subsystem of a certain type of satellite is taken as an example. The information processing subsystem is used for the production of visible light remote sensing data products, including Production Management Subsystem (PMS), Tertiary

Product Production Subsystem (TLPPS), Visible Light Product Production Subsystem (VLPPS), Cataloging Branch System (CBS), File Management Subsystem (FMS). PMS includes Product Planning Decomposition Module (PMS1), Production Order Module (PMS2), Product Production Module (PMS3); TLPPS includes Tertiary Product Production Module (TLPPS1); VLPPS includes Primary Product Production Module (VLPPS1), Secondary Product Production Module (VLPPS2); CBS includes Data Separation Module (CBS1), Geometric Relationship Establishment Module (CBS2), Re-Sampling Module (CBS3), Product Filing Cataloging Module (CBS4); FMS includes Data Extraction Module (FMS1).

In the experiment, according to the template to write the document. And the format and writing of the modes and tasks comply with the specification in SISDL language. Furthermore, the keywords in the document are replaced with corresponding key words in SISDL. Nonstandard codes (such as formulas and task calls) in the document generate corresponding SISDL codes through format specifications. The contents that do not participate in the operation such as names and annotations are removed.

The code format defined by several similar functions obtained from the requirement document through the above steps is shown in Fig. 3.

On this basis, the calling relationship in the system is obtained through the analysis of the requirements document. Figures 4 and 5 respectively show the calling relationship between modes and tasks in the ground information processing subsystem of the satellite. The node pointed by the arrow is the called node, while the independent node does not have a calling relationship.

As can be seen from Fig. 4, PMS calls TLPPS and VLPPS; TLPPS and VLPPS call CBS; FMS is not associated with other parts. According to the system design framework, FMS provides data to other subsystems and has interactions with other subsystems. It can be seen that the FMS part in the requirement document is not perfect.

As can be seen from Fig. 5, the VLPPS1 module, the VLPPS2 module, and the TLPPS1 module call the CBS3 module and the CBS4 module in the product generation process. The PMS3 module calls the VLPPS1 module and the TLPPS1 module; however, the CBS2 module is not called. As can be seen from the original document, it is known that the geometric relationship building module CBS2 needs to be used in the production of products. It can be known that there are errors in the requirement document of VLPPS1 module, VLPPS2 module, and TLPPS1 module. The calling relationship of the task illustrates the relation between the modes in more detail, and also shows the internal structure of the system at the task level.

```
def PMS()
    PPplan=1;
    PRequirement=1;
    PPOrder=1;
PMS1(PPplan);
VLPPS1(PRequirement);
TLPPS1(PPOrder);
end

def CBS()
    InitialData=1;
CBS1(InitialData);
CBS2();
CBS3();
CBS4();
end

def FMS()
    ProductRequest =1;
FMS1(ProductRequest);
end
```

Fig. 3. SISDL code

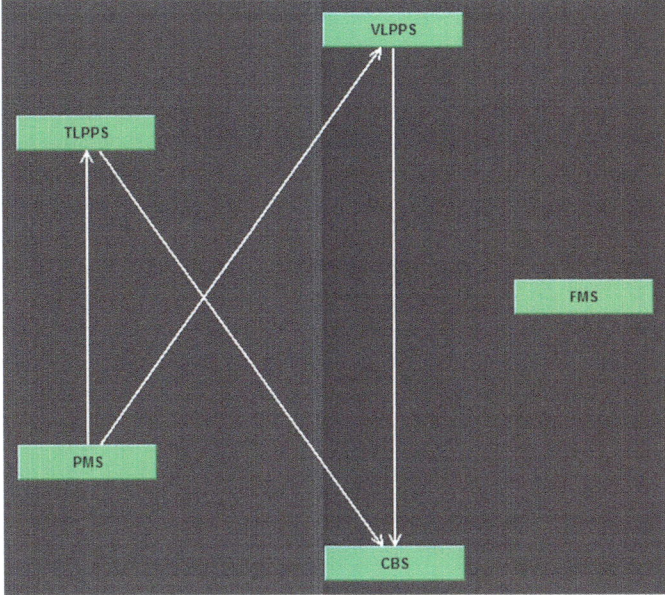

Fig. 4. Mode call

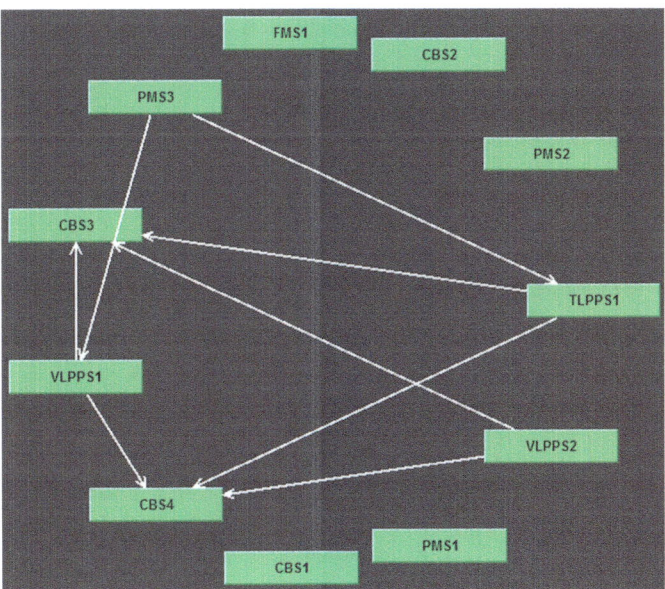

Fig. 5. Task Call

7 Conclusion

In this paper, SISDL description method is designed according to the characteristics of space information systems that focus on modularization and refinement design layer by layer. Requirement Template can assist engineers to write standardized requirement document. The SISDL tool analyzes the requirement document to obtain the requirement model and the structure information of the system. Also, The SISDL tool displays it graphically to help engineers understand the system, check the consistency between the requirement design scheme and the original requirement, and obtain a perfect requirement analysis document for further development work.

Acknowledgements. Fund Project: supported by Gusu Innovation Talent Foundation of Suzhou Under Grant ZXT2017002.

References

1. Wang Zhencao. An Application of System Testing [J]. Information Technology & Standardization, 2007(07):49–51.
2. Dion R. Process improvement and the corporate balance sheet [J]. IEEE software, 1993, 10(4):28–35.
3. Boehm, Turner R. Balancing Agility and Discipline: A Guide for the Perplexed [J]. Lecture Notes in Computer Science, 2004,108(7):316–325.
4. Sheldon F, Kavi K, Tausworth R. Reliability measurement: From theory to practice [J]. IEEE Software, 1992(4):13–20.
5. Meditskos G, Bassiliades N. Structural and Role-Oriented Web Service Discovery with Taxonomies in OWL-S [J]. IEEE Transactions on Knowledge & Data Engineering, 2010, 22(2):278–290.
6. Booch G,Rumbaugh J,Jacobson I.The unified modeling language user guide [M].Beijing: Science Press, 2006.12–26.
7. Lao Huaikou, Chen Yihai. Software formal specification language-Z [M]. Beijing: Tsinghua University Press, 2012.4–11.
8. Abrial J. The B-book: assigning programs to meanings [M]. Qiu Zhongyan Transl. Beijing: Publishing House of Electronics Industry, 2004.427–430.
9. Russo A. Modeling in event-b-system and software engineering by Jean-Raymond Abrial [J]. ACM SIGSOFT Software Engineering Notes, 2011,36(2):38–39.
10. Leveson N. Guest editor's introduction: formal methods in software engineering [J]. IEEE Transactions on Software Engineering, 1990,16(9):929–931.
11. Liu S, Offutt A, Ho-Stuart C. SOFL: A Formal Engineering Methodology for Industrial Applications [J]. IEEE Transactions on Software Engineering, 1998, 24(1):24–45.
12. Liu S. Formal Engineering for Industrial Software Development – An Introduction to the SOFL Specification Language and Method [C]. International Conference on Formal Engineering Methods. Springer Berlin Heidelberg, 2004(3308):7–8.
13. Terence Parr. The Definitive ANTLR4 Reference [M]. Zhang Bo Transl. Beijing: China Machine Press, 2017.3–124.

Design of Routing Incentive Protocol for Space Network Based on Blockchain Technology

Ning Ma[1(✉)], Xi Wang[2], Zhen Li[3], Panpan Zhan[1], Xiongwen He[1], Xiaofeng Zhang[1], and Zhigang Liu[1]

[1] Beijing Institute of Spacecraft System Engineering, Beijing 100094, China
mnuestc@163.com
[2] Beijing Aerospace Linkage and Innovation Technology Co. Ltd., Beijing, China
[3] Institute of Electronics, Chinese Academy of Science, Beijing 100190, China

Abstract. The global commercialization of satellites has intensified, and the types and numbers of new spacecrafts have increased year by year. However, the repeated functional construction and "islandization" characteristics of spacecraft severely restrict the use efficiency of space resources. How to realize the open interconnection and self-adaptation optimization cooperation of satellite resources and encourage more and more on-orbit spacecrafts to share spatial data services and routing resources is a severe challenge for the new generation space network communication protocol. This paper builds the Incentive Protocol of Routing based on the TCP/IP protocol family under the CCSDS spatial communication protocol standard framework. Using blockchain distributed database technology and consensus mechanism, combined with routing flow identification technology, design, and implement a network route incentive protocol and game model based on spacecraft route contribution proof consensus mechanism, intended to provide a feasible technical framework and feasible solution for the construction of the integrated digital intelligent economic ecology of the world.

Keywords: Space networking · Incentive protocol of routing · Blockchain · Space and ground network integration

1 Introduction

With the intensification of the global commercialization of satellites, the types and numbers of new research spacecraft have increased year by year. Under the existing spacecraft development and launch system, the repeated construction of the load function and the isolation feature of "islandization" are widespread, and there is undoubtedly serious restriction and waste on the use of near-Earth space resources. Therefore, in order to realize the open interconnection and adaptive optimization of satellite resources, and encourage more on-orbit spacecraft to share spatial data services and routing resources, higher requirements are put forward for the new generation satellite network communication protocol [1–3].

© Springer Nature Singapore Pte Ltd. 2019
L. Wang et al. (eds.), *Proceedings of the 5th China High Resolution Earth Observation Conference (CHREOC 2018)*, Lecture Notes in Electrical Engineering 552,
https://doi.org/10.1007/978-981-13-6553-9_14

(1) Openness: Any on-orbit spacecraft which belong to different organizations and types can be freely joined or withdrawn from the network without the permission of a third party, subject to the corresponding technical specifications and conditions.

(2) Standardization: A common protocol family architecture should be adopted to maximize the integration of compatible satellite standards and ground Internet communication protocols to achieve integration of space and ground network.

(3) Self-driven: The networking protocol should have a built-in economic model, which can independently optimize the scheduling space routing resources, and encourage spacecraft to share data services: quantify the value of data services and routing resources contributed by in-orbit spacecraft, and Real-time clearing of data services and routing resources provided by the orbiting spacecraft.

(4) Security: Using mature digital signature and encryption algorithm, under the open information transmission system, data encryption transmission and confirmation are realized, and a peer-to-peer security information transmission network is constructed to serve the settlement system.

(5) Neutrality: The space network will serve as a public infrastructure that is not owned and controlled by any single or a small number of institutions and individuals. It will treat the information content of the space network fairly and provide relay and transmission services without review and distinction.

2   Protocol Technology Architecture

In addition to the basic communication functions, the spatial networking protocol quantifies the contribution value of the routing resources (connectivity, transmission data volume, and propagation delay) provided by the spacecraft in the spatial data communication service by the incentive protocol. The identification of the routing information of the space information flow through the spacecraft node is effectively identified and counted, and broadcasted to the entire network in real time. The ground verification node is then responsible for collecting and verifying the broadcast routing information, recording it in the blockchain data book in a distributed competitive accounting manner, and completing the real-time clearing and reward distribution of the spatial network data service and routing resources. The network incentive protocol encourages spacecraft individuals to share data services and routing resources by constructing an economic game model, thereby optimizing the use efficiency of spatial network routing resources.

The incentive protocol architecture is mainly divided into three subsystems: (1) Information Routing Flow Identification Subsystem: completes the routing information identification and statistics of the TCP/IP format exchange distribution information flow between space network spacecraft nodes, and broadcasts to the whole network. (2) Distributed Verification Subsystem: The local verification node checks the broadcast data of the information flow by competing accounting, and stores it in the Distributed File Storage System of IPFS (InterPlanetary File System), and the corresponding IPFS index. The hash value is used for blockchain accounting processing [4].

(3) Self-adaptive optimization scheduling subsystem: The space data service request will divide the real-time priority according to the amount of the additional certificate bonus when it is initiated, and the spatial routing node will give priority to the service of the higher data service request for the certificate. In this subsystem, an economic benefit coordination and game model with positive feedback mechanism is constructed. When each routing node participant adopts the strategic behavior of maximizing its own interests, the utilization efficiency of the entire space network will be optimal (Fig. 1).

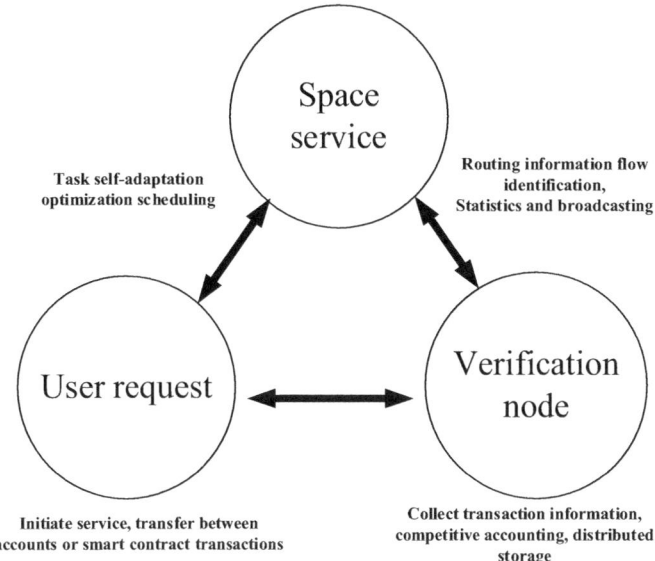

Fig. 1. protocol architecture model diagram

2.1 Information Flow Route Identification

A set of information flows is a set of single-item packets between a given source node and a destination node. Both source and destination nodes are defined by the address of the network layer and the port of the transport layer. The incentive protocol records and analyzes the network traffic information based on the flow network packet exchange technology, distributes the information according to the service request priority, and collects and broadcasts the routing information of each node in real time. This process is divided into two steps

(1) Data exchange and distribution according to the optimal use efficiency of the routing performance of the own node and the priority order of the transfer target data. The prioritization can be expressed as the amount of the token attached that the user is willing to pay when requesting the service.

(2) After the data transmission is completed, the distribution data stream is collected and broadcast to the entire network. The broadcast statistics include: the incentive layer protocol version number, the source port IP address, the source port ID (public key address), the target port IP, Target port ID (public key address), data transfer amount, transfer start and end time stamp, and stream ID.

2.2 Distributed Consensus Synchronization

The distributed consensus subsystem is mainly responsible for verifying and storing the information flow identification data of the broadcast, and forming the consensus of the whole network and recording it in the blockchain ledger. Considering the operating costs, the verification node body is generally composed of a ground network computer. After the verification node verifies the routing information broadcast by the space routing incentive protocol, the route contribution value of each spacecraft routing node in the data service is converted, packaged, and stored in the IPFS (InterPlanetary File System, IPFS) distribution. In the data sharing system, the index hash address of the packet in the IPFS is stored in the blockchain data book as the workload of the routing node to obtain the certificate revenue.

The information packed by the verification node includes not limited to the following contents: (1) digital signature of the source node, the destination node and the relay node in the transmission service; (2) transmission time stamp of each node; (3) data transmission amount; (4) The data packet hash value of each node; (5) The verification result of the data service correctness of the destination node at the time of service termination. The process of verifying the routing of the routing information by the node is as follows (Fig. 2).

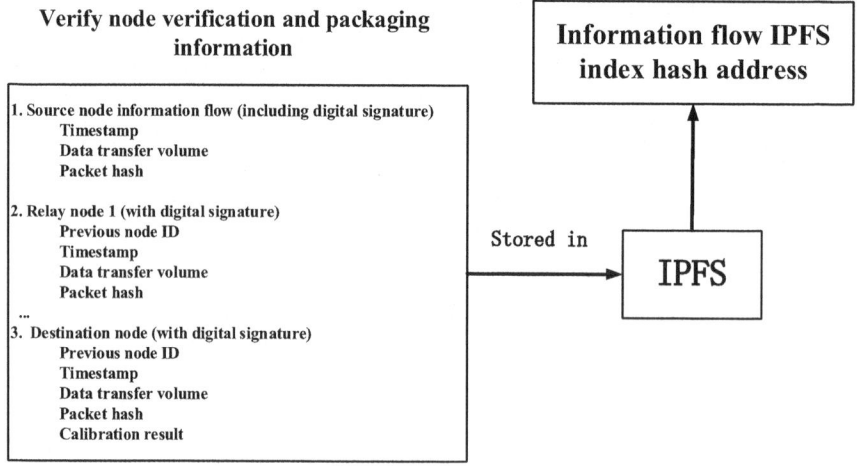

Fig. 2. Information flow IPFS index hash generation diagram

Considering that it is expensive to store data for blockchain, only the IPFS index hash address corresponding to the key data is stored in the blockchain ledger, and the complete data content is stored in the IPFS system. The IPFS system is based on a globally distributed peer-to-peer encryption storage protocol. Any node can only store one fragmented content of a complete file, and each fragment is backed up in multiple IPFS storage nodes, and only the complete index hash address is obtained [5, 6]. In order to collect all the file fragments and extract the complete file. Therefore, the scheme design of adopting such index address winding can not only ensure the security of data storage, but also save construction cost and improve the running efficiency of the entire system.

The generation of a block will package the following information: (1) The hash address of the collected and verified information flow during the period; (2) The transaction information between the collected and verified accounts during the period; (3) Block Generate timestamp information; (4) Incentive protocol version number; (5) Block height; (6) Previous block hash value. Therefore, the ground verification node is a non-tamperable high-performance chained distributed storage database [7, 8] (Fig. 3).

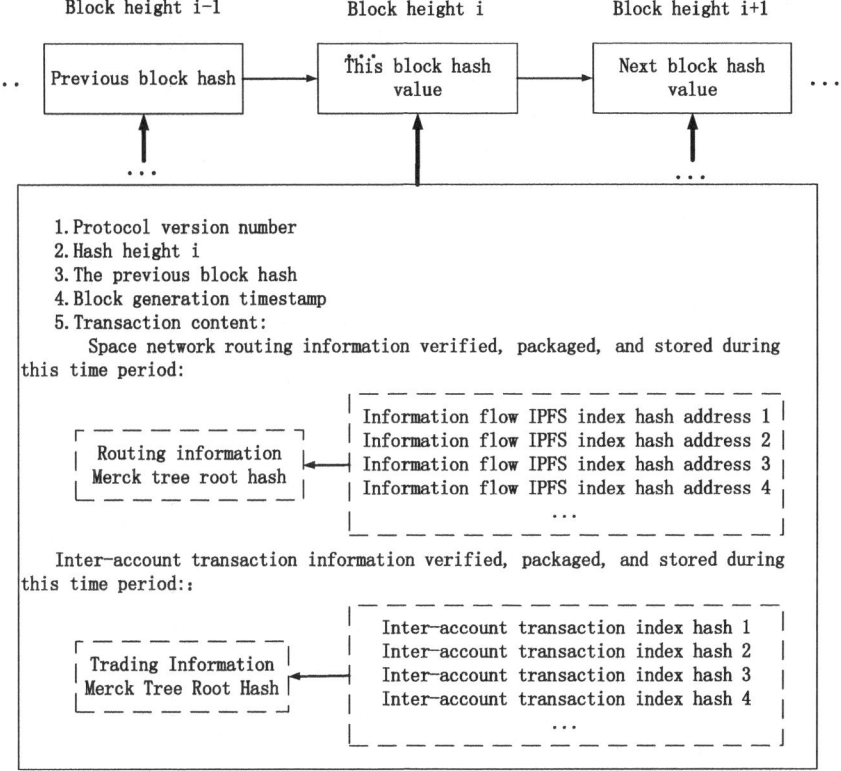

Fig. 3. block structure diagram

2.3   Self-adaptive Optimization Scheduling

The implementation of the self-adaptive optimization scheduling subsystem relies on constructing a credible economic model that can coordinate the interests of all parties and has a positive feedback incentive effect. The incentive agreement involves three parties: the space routing node, the ground verification node, and the user. The constructor of the spatial network service is mainly composed of the spatial routing node and the verification node: the spatial routing node forms a spatial network to provide spatial routing resources and data services for users; the ground verification node collects and verifies the routing contribution of routing nodes in the spatial data service. Value and inter-account transaction information, store key information and generate a blockchain transaction book that cannot be tampered with. The user holds a certain amount of pass and enjoys a certain share of the space network, and can use the pass to purchase data services or increase the priority of service requests.

The implementation of the self-adaptive optimization scheduling subsystem relies on constructing a credible economic model that can coordinate the interests of all parties and has a positive feedback incentive effect. The incentive agreement involves three parties: the space routing node, the ground verification node, and the user. The constructor of the spatial network service is mainly composed of the spatial routing node and the verification node: the spatial routing node forms a spatial network to provide spatial routing resources and data services for users; the ground verification node collects and verifies the routing contribution of routing nodes in the spatial data service. Value and inter-account transaction information, store key information and generate a blockchain transaction book that cannot be tampered with. The user holds a certain amount of pass and enjoys a certain share of the space network, and can use the pass to purchase data services or increase the priority of service requests (Fig. 4).

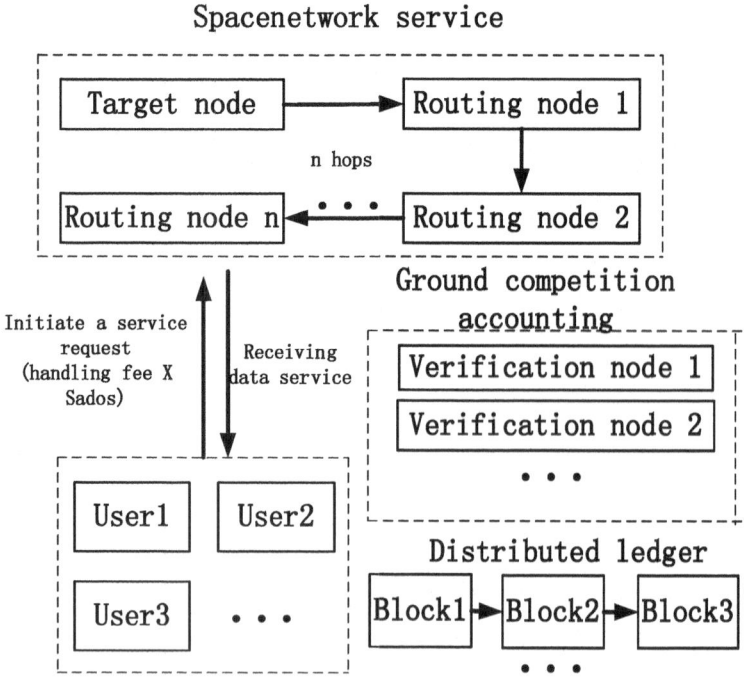

Fig. 4. Schematic diagram of self-adaptive optimization scheduling

The economic incentive model is designed as follows: The pass name used to measure the space data service of the whole network is the Satdomain, shorthand: Sado, which characterizes the holder of the Sado to enjoy the use of the routing resources of the space network. The initial circulation of Sado is 20,000,000, the final circulation is 200,000,000, the full issuance period is 45–106 years, and the annual issuance limit is 2–5%. The annual Sado issuance rate is based on the growth rate of the total service value of the previous year. The full issuance period is 45–106 years. The additional issuance amount is used to reward the contribution of the space routing node and the verification node as the infrastructure.

The following economic indicators are defined for the entire network:

(1) $GSSP_j$: Gross Spacenetwork Service Product AD j.

$$GSSP_j = \sum_{i=1}^{mj} Hop_{ij} * MB_{ij} \tag{1}$$

m_j The total number of information flow routing services provided by the whole network in AD j.
Hop_{ij} the number of hops of the i-th space routing data service in AD j.
MB_{ij} The data transmission volume of the i-th space routing data service in AD j.

(2) Pj: The growth rate of GSSP in the previous year compared to the previous year in AD j.
(3) Sj: The total amount of Sado basic certificate issued in AD j.
(4) ΔSj: New circulation of Sado in AD j.

The following design for the entire network Sado distribution and rewards.

The Sado characterizes the use of space network data services. In order to maintain the stability of the Sado currency, the increase in the total amount of Sado's issuance should always be consistent with the growth of GSSP (Gross spacenetwork service production). The total value of the default space network service will reach saturation in 106 years, that is, the Sado circulation will reach no increase after the number of 200 million issuance. During this issuance, the annual new circulation of Sado is always issued according to ΔSj (when ΔSj meets the conditions: $2\% < \Delta Sj < 5\%$), the current year $GSSP_{j-1}$ growth rate ΔP_{j-1} is also in the range of 2–5%, then in the current year Sado new circulation

$$\Delta S_j = \Delta P_{j-1} * S_{j-1} \tag{2}$$

The increase in the total value of the space network service is due to the expansion of the scale of the service provider's space routing node and the verification node and the improvement of the efficiency. Therefore, the issued Sado is used for proportional rewards and routing nodes and verification nodes. 95% of the new circulation is awarded to the routing node. 5% of the new circulation is awarded to the verification node. In addition to this, the routing node and the verification node will also obtain a transaction reward for the user's additional payment.

Annual total revenue distribution of space routing nodes in AD j:

$$E_{Rj} = X\frac{n}{n+1} + 95\% * \Delta S_{j-1} \tag{3}$$

X   Annual user additional payment fee;
n   average hop count of the annual information flow space route;

Annual total revenue distribution of the ground verification node in AD J

$$E_{Vi} = X\frac{1}{n+1} + X_{j-1} + 5\% * \Delta S_{j-1} \tag{4}$$

For a single space routing node A and a verification node B, the respective proportion of the total annual route contribution and the verification contribution is k_{ar} and k_{bv} then

Annual total revenue of space routing node a in AD J

$$E_R = \sum_{i=1}^{a_j} \frac{Xa_i}{n_{ai}+1} + 95\% * k_{ar} * \Delta S_{j-1} \tag{5}$$

a_j the total number of routing data services in the space routing node A in AD j;
X_{ai} the additional payment fee for routing node A's i-th routing data service;
n_{ai} the average hop count of routing node A's i-th routing data service space;

The total annual revenue of the ground verification node b in AD J:

$$E_V = \sum_{i=1}^{b_j} \frac{X_{bi}}{n_{bi}+1} + 5\% * k_{bv} * \Delta S_{j-1} + X_{bj} \tag{6}$$

b_j the total number of data services verified by ground verification node B in AD j.
X_{bi} the additional payment for ground verification node B's i-th verifing data service;
n_{bi} the average hop count of ground verification node B's i-th verifing data service space;
X_{bj} The fee charged by the ground verification node B in the transaction service in AD j.

3   Application Prospects and Value—Space and Ground Integration Intelligent Digital Ecological Economy

This paper designs the technical framework of spatial networking route incentive protocol based on blockchain technology, and demonstrates that the economic significance of its certificate-exciting model plays an important role in promoting the construction of the worldwide integrated network. At the same time, the technical framework of the incentive agreement has a high degree of expansion and broad application space. Through the incentive agreement, the economic model is conducive to building a worldwide intelligent network economy based on spatial data services, creating a fair, mutually beneficial and diversified intelligent contract platform and network intelligent economic ecology.

References

1. Satellite Industry Association. State of the satellite industry report[EB/OL]. [2016-03-02]. http://www.sia.org/wpcontent/uploads/2013/06/2013_SSIR_Final.pdf.
2. Iridium. Iridium NEXT: The bold future of satellite communications [EB/OL]. [2016-03-02]. https://www.iridium.com/network/iridiumnext.
3. Wang Jiasheng. China data relay satellite system and its application prospect[J]. Spacecraft Engineering, 2013(1): 1–6.
4. IPFS - Content Addressed,Versioned, P2P File System[EB/ OL]. https://blog.acolyer.org.
5. I.BitTorrent. Bittorrent and ˆA¸ttorrent software surpass 150 million user milestone, Jan. 2012 [Z].
6. L. Wang and J. Kangasharju.Measuring large-scale distributed systems : case of bittorrent mainline dht[C]. In Peer-to-Peer Computing (P2P), 2013 IEEE Thirteenth International Conference on, IEEE, 2013:1–10.
7. Mukhopadhyay U, Skjellum A, Hambolu O, Oakley J, Yu L, Brooks R. A brief survey of Cryptocurrency systems. In: Privacy, Security and Trust. 2017. 745–752.
8. Antonopoulos AM. Mastering Bitcoin: Unlocking Digital Crypto-Currencies. O'Reilly Media, Inc., 2014.

Real-Time GF-4 Satellite Image Enhancement in Low Light Conditions

Mingming Ma, Bojia Guo, and Yi Niu[✉]

School of Artificial Intelligence, Xidian University, Xi'an, China
niuyi@mail.xidian.edu.cn

Abstract. GF-4 satellite images suffer from a general drawback of geostationary satellites that lose the efficacy on low light conditions such as gloaming and drawn. This significantly limits the GF-4 applications when disaster happens. In this paper, we investigate a real-time image enhancement technique for GF-4 satellites images on low light conditions and propose a reflectivity-constraint-entropy-maximization (RCEM) strategy to enhance the contrast of GF-4 images. Experiments show that the proposed RCEM outperforms the current state-of-the-art image enhancement techniques which will significantly extend the working time of GF-4 satellite.

Keywords: GF-4 satellite · Image enhancement · Real time · Dynamic programming

1 Introduction

GF-4 satellites are the highest resolution geostationary satellite of China. Taking advantages of its large fields and high spatial and temporal resolution, it significantly improves the application of Chinese disaster prevention, forestry surveillance, etc., applications. However, it suffers from the general drawback of geostationary satellites that the working periods is highly limited by the light conditions, for example, at drawn and gloaming. This affects the real-time reaction when disaster occurs. Therefore, in this paper, we investigate a real-time image enhancement technique for the GF-4 images at low light conditions. Current image enhancement techniques can be roughly classified into two types: global operations and local operations. The main idea of local operators is to adopt diverse kinds of filters to modify the specific frequency components of the input images. Such as adopting high-pass filters to enhance the details like sharp edges and textures, or adopting homomorphic filters or bilateral filtering [1] to suppress the low frequency components, because the low frequency components can be regarded as the illuminance of the scene. Using these kinds of techniques can fix the low contrast image caused by large illuminance difference, such as taking pictures at night under a road lamp. There is another kind of local operation techniques which uses filters to simulate the human visual system. These techniques use a compound word

© Springer Nature Singapore Pte Ltd. 2019
L. Wang et al. (eds.), *Proceedings of the 5th China High Resolution Earth Observation Conference (CHREOC 2018)*, Lecture Notes in Electrical Engineering 552,
https://doi.org/10.1007/978-981-13-6553-9_15

"retinex" which is a combination of "retina" and "cortex" [1, 2]. To be specific, retinex techniques use a set of Gaussian filters with different variances to simulate the collaboration working mechanism of retina and cortex in human visual system.

Although the local operation techniques can significantly improve the contrast of image textures, it is improper for the satellite image enhancements, because the local operations will break the rank of pixel intensity levels. This seems acceptable for consumer electronic applications, but at the context of satellite images. The final objection of image enhancement is for a more "visual pleasing image", instead, the intensity of pixel values reflects the reflectivity of the scene. Changing the rank may causes imponderable results on the disaster prevention and suppression. For example, when flood occurs, the experts have to monitor the regions of flood and make real-time decisions for flood discharge. Changing the rank of pixel values may cause the common "halo" artifacts and misleading the experts of the flood conditions.

Therefore, we investigate the global image enhancement techniques for GF-4 image enhancement tasks. The global image enhancement techniques use one unique transfer function to update the intensity levels into new pixel values. Since all the pixels adopt a unique transfer function, the intensity level rank can be preserved. Another advantage of using global transfer function is that the complexity of the enhancement technique is only related to the histogram instead of the image size. Considering the high resolution of GF-4 images ($10k \times 10k$), it will favor the real-time implementation. The most widely adopted transfer function is the series of histogram equalization techniques [3–5]. In [4], the author proposed to constrain local histogram equalization technique via balancing two conflicting requirements: the maintenance of the image average luminance appearance and the contrast enhancement of the details. In [5], the author proposed a histogram adjusting technique which investigates an optimum fusion strategy between the classic histogram equalization operator and linear mapping operator.

However, the current global image enhancement techniques suffer from two main drawbacks which are local optimum solution and unconstraint L_∞ reflectivity distortions. The local optimum problem is caused by the cumulative density function based optimization technique. Take HE as an example, the transfer function is calculated as

$$\hat{I}(k) = \text{floor}\left(\sum_0^k P_i \times k \right) \tag{1}$$

where k is the input pixel value, P_i is the probability of k. It can be observed from Eq. (1) that the $\hat{I}(k)$ only depends on the cumulative density function P_i, and determined sequentially. According to the optimization theory, the cumulative density function based technique can be regarded as using the greedy search strategy to achieve the maximum entropy. For example, if the input histogram is $(0.3, 0.1, 0.2, 0.4)$, the global optimum transfer function should combine the second and third bin of the histogram and generate $(0.3, 0, 0.3, 0.4)$ (Fig. 1).

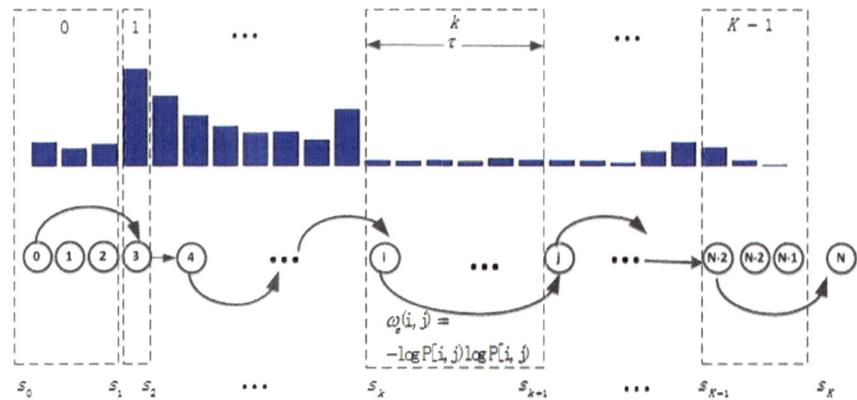

Fig. 1. The graph representation of the proposed RCEM technique

But with Eq. (1), the transfer function will combine the first and second bin and generate a histogram with $(0.4, 0, 0.2, 0.4)$. The local optimal problem will affect the exhibition of local details of the GF-4 images.

The unconstraint L_∞ reflectivity distortions problem is caused by the non-constraints objective functions of the current global operation techniques. Since there are no constraints on Eq. (1), the transfer function is prone to combine all the adjacent intensity values of the histogram. It seems fine for general image enhancement tasks, but for GF-4 image visualization tasks, the adjacent intensity values represent the subtle variance of local objects in the scene. Combining too many adjacent bins may cause the loose of discriminative reflectivity on target intensity ranges.

Therefore, in this paper, we propose a reflectivity-constraint-entropy-maximization (RCEM) technique to overcome the above two drawbacks. The objective function of RCEM is similar to traditional HE technique which maximizing the entropy of the GF-4 image, but with a constraint to bound the maximal bins that can be combined. In addition, we propose a dynamic programming based optimization technique to solve the constraint-entropy-maximization problem that the global optimum solution can be achieved for sure. Experimental results show that, the proposed RCEM technique outperforms the current global image enhancement techniques and in exhibition more details will preserve the subtle variance.

2   Reflectivity-Constraint-Entropy-Maximization

2.1   RCEM Formulation

The input of the RCME technique is the histogram of the GF-4 image I; Let N to be the number of bins of GF-4 image, $N = 1024$ for the first five visible bands of GF-4 image and $N = 4096$ for the infra-red band of GF-4 images, K determines the number of output bins, $K = 256$ for common settings. First of all, please allow us to give a mathematic definition of a intensity value transform function T via an ordered vector $\mathbf{s} = (s_1, s_2 \ldots s_{K-1})$ with integers such that all input intensity values i between the interval $[s_K, s_{K+1})$ are modified into output pixel value k, written as $T(s_k, s_{k+1}) = k$, $0 \leq k < K$, where $s_0 \equiv 0, s_K \equiv N$. The interval size

$$\varepsilon_k = s_{k+1} - s_k \tag{2}$$

is actually the input bins that are combined to the output value k. In this way, the interval ε_k can be regarded as the reflectivity distortion caused by image enhancement. Therefore, by bounding ε_k RCEM can pretend too many bins of the input histogram to be mapped into one single output value, which provides the subtle variance in smooth regions and avoids distorting reflectivity continuity. Let $P[i, j]$ to be the cumulative probability that the intensity values in I is in the range $[i, j)$. Then, the image enhancement tasks can be modeled as a constrained RCEM optimization problem as follows:

$$\max_s \sum_0^{N-1} -P[s_K, s_{K+1}] \log P[s_k, s_{k+1}] \tag{3}$$

$$\text{subject to } s_{k+1} - s_k \leq \tau \; \forall k$$

where τ is the upper bound reflectivity distortion which limits the maximum bins that can be combined in the enhancement process.

A straightforward way to solve the problem of Eq. (3) is using the cumulative density function based techniques like HE. However, as we discussed before, the cumulative function based technique suffers from the local optimum problem. Therefore, we investigate a global optimum solution for Eq. (3). We resort to the graph theory and model the input GF-4 histogram as a directed acyclic graph (DAG) $G(V, E)$. Every intensity values corresponds to the vertices V of $G(V, E)$, and the edges $e[i, j]$ between vertex v_i and v_j denotes the penalty of combining the intensity range $[i, j)$ to one output bin. In this way, Eq. (3) can be formulated as a maximum-weight path problem with K-edges of an directed acyclic graph (DAG) if $e[i, j]$ is assigned a weight.

$$w_e(i, j) = -P[i, j] \log P[i, j] \tag{4}$$

To construct the above-mentioned DAG $G(V, E)$, we meet the upper bound constraints on reflectivity distortion by eliminate all the edges which connect vertex i to vertex j if $j - i \geq \tau$. In this way, the optimization problem (3) can be converted into searching a path with K-edges that start from vertex 0 and end at vertex N. The weight of the path is defined by the sum of the edge weights in this path. Now we are seeking for the path with maximal weights. In this way, Eq. (3) can be written as the following K-edges maximum-weight as:

$$\hat{L}(N, k) = \arg\max_s \sum_{k=0}^{K} w_e(s_k, s_{k+1}) \tag{5}$$

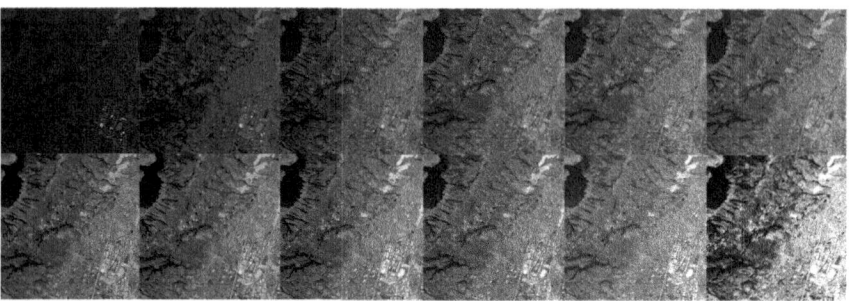

Fig. 2. The comparisons of the proposed RCEM methods with different τ, from top-left to bottom-right $\tau = (4, 6, 8, 12, 14, 16, 18, 20, 40, 60, 80, 256)$

2.2 Dynamic Programming Solution

Given the definition of the intensity value transform operator T via an integer vector $\mathbf{s} = (s_1, s_2 \ldots s_{K-1})$, there are C_N^K possible solutions. It is intractable to enumerate all the possible solutions and find the optimal one. Fortunately, the optimization problem (5) can be recursively divided into several sub-problems and calculated via iteration

$$
\begin{aligned}
\hat{L}(N, k) &= \arg\max_s \sum_{k=0}^{K} w_e(s_k, s_{k+1}) + w_e(s_{K-1}, s_K) \\
&= \arg\max_{s_{K-1}} \hat{L}(Ns_{K-1}, K-1) + w_e(s_{K-1}, s_K)
\end{aligned} \tag{6}
$$

Actually, in the discrete mathematics [6], (6) can be solved via dynamic programming which is more memory and time efficient than iteration solution. Therefore, we investigate a dynamic programming based technique to solve (6), and illustrate the pseudo-code in Algorithm 1.

2.3 Complexity Analysis

Now we analyze the complexity of the proposed RCEM technique. RCEM consists of three steps: (1) histogram calculation, (2) graph generation, and (3) dynamic programming. The histogram calculation is a general step for all the global image enhancement techniques which are linear to the image size and can be ignorable. So we only analyze the complexity of (2) and (3).

For the graph generation, the default time complexity is $O(N^2)$ via traversing all the combination of nodes, but considering the reflectivity constraint, the time complexity of graph generation is only $O(N\tau)$. The most time-consuming part of RCEM is the dynamic programming. From Algorithm 1, we can observe that the dynamic programming has to fill two tables $\mathbf{W}$ and $\mathbf{R}$. the size of $\mathbf{W}$ and $\mathbf{R}$ is $N \times K$. Filling each cell has to traverse τ possible candidates. Therefore, the time complexity of dynamic programming is $O(\tau NK)$. The overall complexity $O(NK\tau)$ which is linear to the maximum pixel value N. Since for GF-4 visible band images, $N \leq 1024$, it can be calculated in real time (less than 0.1 s).

Algorithm 1: Pseudo-code of the dynamic programming solution of RCEM

Input: The N bins of the input GF-4 histogram.
   The maximum reflectivity distortion constraint τ
Output: The transfer function T defined by a integer vector $\{ \mathbf{s} = (s_1, s_2 \dots s_{K-1}) \}$

STEP1: Graph construction
for each i **less than** N **do**
   for each j **less than** τ **do**
      $w_e(i, j) = -P[i, j] \log P[i, j]$
   end
end
STEP2: Intimidate results Initialization
1. A $N \times K$ matrix $\mathbf{W}$ for the sub-problems $\hat{L}(n, k)$.
2. A $N \times K$ matrix $\mathbf{R}$ for the searching back of cutting point s_{K-1} of $\hat{L}(n, k)$.
for each n **less than** N **do**
   $\mathbf{W}(n, 1) = w_e(0, n)$
   $\mathbf{R}(n, 1) = 0$
end
STEP3: Dynamic programming process
for each k **in range** *[1,K]* **do**
   for each n **in range** $[k, k + \tau]$ **do**
      get $\hat{L}(n, k)$ and s_{K-1} via (6);
      $\mathbf{W}(n, k) = \hat{L}(n, k)$
      $\mathbf{R}(n; k) = s_{K-1}$
   end
end
STEP4: Searching back of $\mathbf{R}$ to get the transfer function T
$s_K = N$
for k **in range** *[K-1, -1,0]* **do**
   $s_K = \mathbf{R}(s_{K+1}, k)$
end
return $\mathbf{s} = (s_1, s_2 \dots s_{K-1})$

---

3 Experiments

In this section, we conduct experiments for RCEM from two aspects. First, we verify the settings of τ. Then we evaluate the perceptual quality of RCEM method in comparison with Linear Rescaling (LR) and Histogram Equalization (HE). The test image is from GF-4 visible light bands with $N = 1024$, $K = 256$.

The first task is to set the reflectivity distortion constraint. To evaluate the performance of different τ, we tune τ and illustrate the results in Fig. 2. When $\tau = 4$, the RCEM technique forces to find a path connecting 0 and 1024 with 256 edges, and the maximal length of each edge is 4. In this way, RCEM technique becomes a linear mapping operation which achieves a dark image as shown in the top left. With the increase of τ, the contrast of the image became more obvious but reflectivity distortion arises. When ($\tau = N$), the RCEM technique became a non-constraint technique which achieves an image like histogram equalization (denoted by HE) and incurs the largest reflectivity distortions. From Fig. 2, we can conclude that $\tau = 14$ makes a good balance between high entropy (good contrast) and low reflectivity distortion.

Then we compare the perceptual quality. Figures 3, 4, 5 and 6 exhibit the visualized low dynamic range image produced by linear mapping (denoted by LR). For the reason that the images in low light conditions are linearly mapped to high dynamic range images, there are no details enhanced in the results, including small targets, artificial buildings, and other regions of interest. HE is a well-known method for image enhancement, but it may easily over-enhance the images in low light condition as Fig. 3b shows, in which details in the images are easily loosed. Compared with LR and HE techniques, it can be seen that the proposed RCEM method catches the best visual effect as Fig. 3c illustrates. RCEM prevents small targets to be compressed to a single bin by maximize the entropy to a large extent as Fig. 4c shows. And it provides more clear and identifiable edges and shape for targets like streets, small houses, etc. That is extreme important for satellite applications such as recognition tasks.

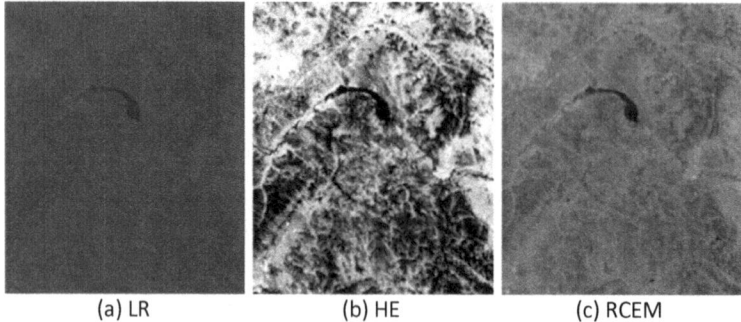

(a) LR   (b) HE   (c) RCEM

Fig. 3. Comparisons of visualized GF-4 images in low light condition by different enhancement techniques

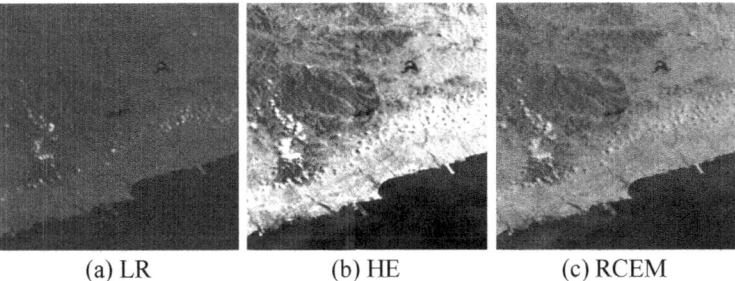

(a) LR   (b) HE   (c) RCEM

Fig. 4. Comparisons of visualized GF-4 images in low light condition by different enhancement techniques

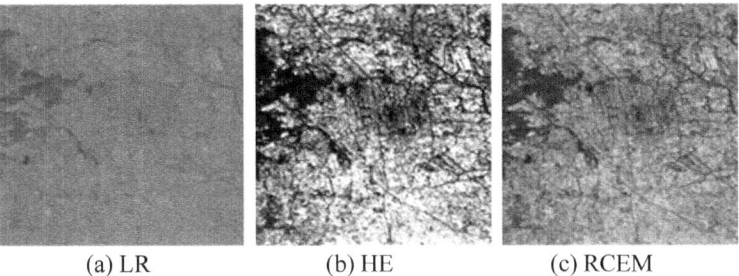

(a) LR   (b) HE   (c) RCEM

Fig. 5. Comparisons of visualized GF-4 images in low light condition by different enhancement techniques

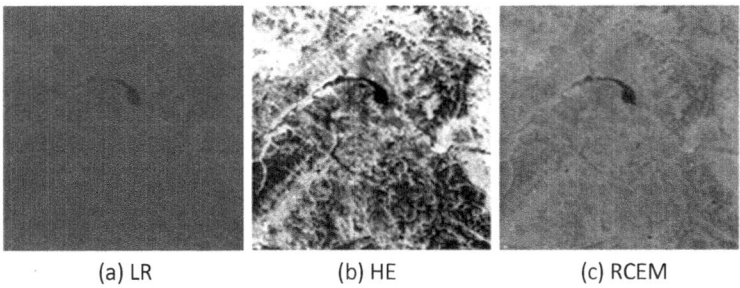

(a) LR   (b) HE   (c) RCEM

Fig. 6. Comparisons of visualized GF-4 images in low light condition by different enhancement techniques

4 Conclusion

This paper proposes a real-time image enhancement technique via reflectivity-constraint-entropy-maximization (RCEM) for GF-4 satellite images on low light conditions. RCEM preserves the rank intensity and can maximize information exhibition theoretically while constraining reflectivity error bound. The performance of RCEM is can be adjusted by the only parameter τ in different applications, and all the experiments show that RCEM outperforms LR and HE techniques with more clear details.

References

1. C. Tomasi and R. Manduchi, "Bilateral filtering for gray and color images," in *Proceedings of the Sixth International Conference on Computer Vision*, ser. ICCV '98. Washington, DC, USA:IEEE Computer Society, 1998.
2. Z.-u. Rahman, D. Jobson, and G. Woodell, "Multi-scale retinex for color image enhancement," in *Image Processing, 1996. Proceedings., International Conference on*, vol. 3, Sep 1996, pp. 1003–1006 vol. 3.
3. S. M. Pizer, E. P. Amburn, J. D. Austin, R. Cromartie, A. Geselowitz, T. Greer, B. ter Haar Romeny, J. B. Zimmerman, and K. Zuiderveld, "Adaptive histogram equalization and its variations," *Computer Vision,Graphics, and Image Processing*, vol. 39, no. 3, pp. 355–368, 1987.
4. H. Zhu, F. H. Chan, and F. Lam, "Image contrast enhancement by constrained local histogram equalization," *Computer Vision and Image Understanding*, vol. 73, no. 2, pp. 281–290, 1999.
5. J. Duan, M. Bressan, C. Dance, and G. Qiu, "Tone-mapping high dynamic range images by novel histogram adjustment," Pattern Recognition, vol. 43, no. 5, pp. 1847–1862, 2010.
6. T. H. Cormen, C. E. Leiserson, R. L. Rivest, and C. Stein, *Introduction to Algorithms*. MIT Press, 2009.

Development and Prospect of Spaceborne Infrared Fourier Spectrometer

Ren Chen[1,2], Cong Gao[1,2,3(✉)], and Jianwen Hua[1,2]

[1] Shanghai Institute of Technical Physics, Chinese Academy of Sciences,
Shanghai 200083, China
gaocong92@163.com
[2] Key Laboratory of Infrared Detection and Imaging Technology, Chinese
Academy of Sciences, Shanghai 200083, China
[3] University of Chinese Academy of Sciences, Beijing 10049, China

Abstract. The satellite-borne infrared Fourier spectrometer is one of the main infrared remote sensing instruments mounted on meteorological satellites. It can detect the vertical distribution of parameters such as large temperature and humidity. In this paper, the development of spaceborne infrared Fourier spectrometer is reviewed, and the performance characteristics of the spaceborne infrared Fourier spectrometer at a different stage at home and abroad are compared. Combined with the development of current instruments and the characteristics of the information age, the process of spaceborne infrared Fourier spectrometer technology development is prospected, and the development direction of "one wide and three high" and hyperspectral integrated data network is pointed out, which can provide reference for maturity and practicability of the infrared Fourier spectrometer technology in China.

Keywords: Fourier transform spectrometer · Infrared hyperspectral detection · Meteorological observation

CLC number: V447+.1

1 Introduction

Satellite-borne infrared Fourier spectrometer is one of the main infrared remote sensing instruments carried by Meteorological satellites. It can measure the specific infrared spectrum of the earth's atmosphere continuously and accurately from space by using the infrared radiation properties of atmosphere and earth surface. Though Fourier transform of raw data, the vertical fraction of the parameters such as air temperature and humidity can be calculated. Combined with the data collected by the instrument at various points on the earth, we can obtain the urgently needed three-dimensional information of atmospheric elements for meteorological application, and realize large-scale, rapid, continuous and long-term meteorological measurement. This is considered to be the most significant achievement in the history of meteorological observation since the development of radio sounding technology.

© Springer Nature Singapore Pte Ltd. 2019
L. Wang et al. (eds.), *Proceedings of the 5th China High Resolution Earth Observation Conference (CHREOC 2018)*, Lecture Notes in Electrical Engineering 552,
https://doi.org/10.1007/978-981-13-6553-9_16

Satellite-borne infrared Fourier spectrometer has been successfully used in the fields of atmospheric detection, gas composition detection, and so on. It has played a great role in the development of meteorological detection and other fields such as national security and economic construction which has not yet been paid attention to. With the successful launch of the atmospheric vertical detector on FY-4, the attention to infrared Fourier hyperspectral detection technology in the meteorological field has reached an unprecedented height, which brings new opportunities and challenges of the research work on spaceborne infrared Fourier spectrometer in China. In this paper, the characteristics of spaceborne infrared Fourier spectrometer technology are introduced, the development history is reviewed, the present status is summarized, and the trend in the future is discussed and analyzed.

2   Technology Development of Infrared Fourier Spectrometer Abroad

2.1   CrIS [1–4]

CrIS is one of the important payloads of the NPOESS system. It was launched successfully in 2011. It measures the infrared radiation spectrum of the upwelling earth-atmosphere with high spectral resolution and high radiation accuracy. The data obtained and the microwave remote sensing data of the NPOESS platform are used to construct the high-precision vertical profiles of the earth's atmospheric temperature, humidity, and pressure.

CrIS and Microwave Detector (CMIS) are collectively known as the CRIMSS. The data produced by it related to high-precision vertical profiles of atmospheric temperature, humidity, and pressure are a set of basic input parameters of the numerical weather prediction model. It will greatly improve the accuracy of weather prediction of weather models, storm tracks and precipitation in the global scope by establishing a reasonable meteorological model.

CrIS is a three-band Michelson interferometer. Each band has a circular sensing aperture of a 3 * 3 array on its focal plane. The optical design includes a foldable Grigori telescope located behind the interferometer, a field-of-view aperture that limits the sensor aperture array and an optical system collecting the interferogram energy onto a photovoltaic HgCdTe detector. The key points of optical design include: telescope structure, pupil position, channel spectral elimination, polarization sensitivity, stray energy suppression, miniaturization design, etc.

CrIS is the most representative infrared atmospheric sounding instrument for Polar Orbiting Meteorological satellites.

Its main technical indicators are shown in Table 1.

The optical model is shown in Fig. 1.

Table 1. CrIS technical indicators

Optical aperture	8 cm		
Orbit height	833 km		
Spectral range	650–1095 cm^{-1}	1210–1750 cm^{-1}	2155–2550 cm^{-1}
Spectral resolution	0.625 cm^{-1}	1.25 cm^{-1}	2.5 cm^{-1}
Optical field of view	3.3° × 3.1°		
Spatial resolution	14 km		
Detector element number	3 × 3		
Volume	80 cm × 47 cm × 56 cm		
Weight	<98 kg		
Power	<95 W		

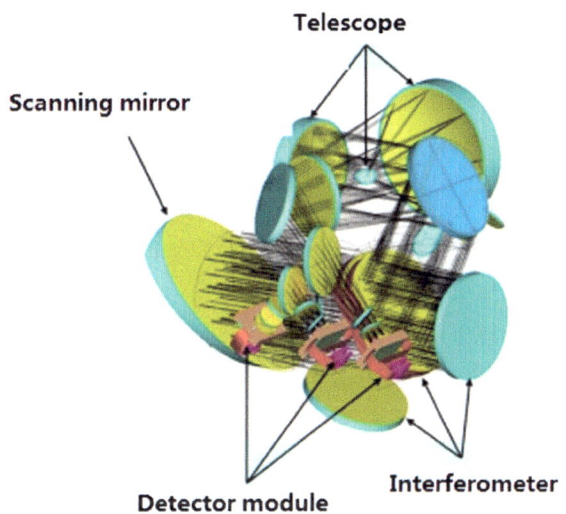

Fig. 1. CrIS model diagram of optical system

2.2 IASI [5]

The IASI is a new type of atmospheric vertical sounder based on the Michelson interferometer (Fig. 2), which has been studied jointly by the French and Italian space research institutes since 1990. It is an important new type of load on the MetOp satellite (European Polar Orbit Operational Meteorological Satellite). The IASI can provide thousands of infrared information channels, greatly increasing temperature and humidity. The vertical resolution is very good to meet the needs of the World Meteorological Organization. The first IASI flight prototype was launched on October 19, 2006.

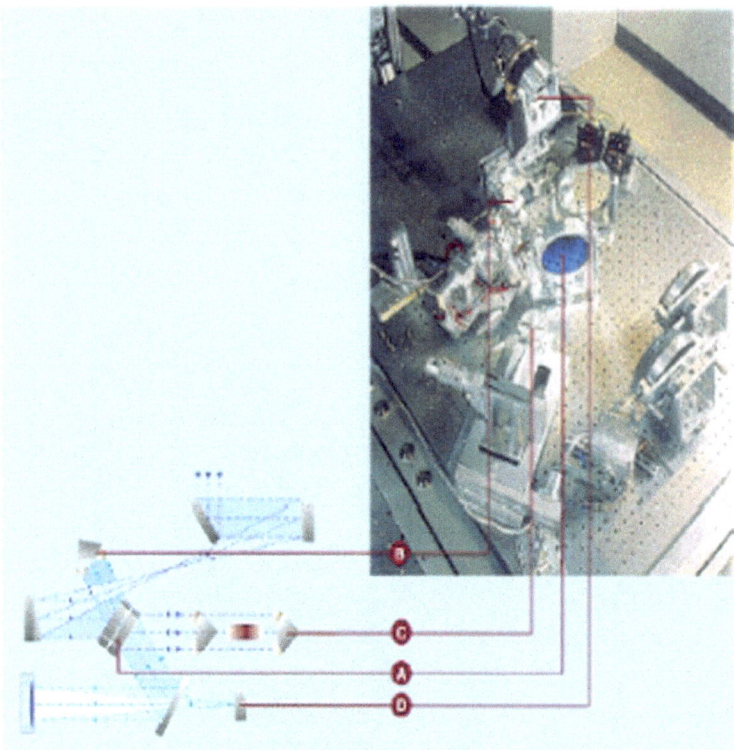

Fig. 2. IASI partial optical road map (A beam splitter B static mirror C Dynamic mirror D cold box system)

The mirror of IASI interferometer is a hollow three-dimensional angular mirror. The moving mirror has a moving distance of ± 2 cm and its working temperature is 300 ± 2 K. The field diaphragm, aperture diaphragm, spectral separation light path, and detector are all placed in the cold box (Table 2), the temperature of the cold box is lower than 100 K (about 93 K). Three-stage passive refrigerators are used for refrigeration. The refrigerators also have the ability to heat up to 60 °C for decontamination purposes. In addition, the sun shocks are designed to prevent the sun from irradiating the refrigerators and affect the effect of low temperature. The interior of the sun cover is plated and polished.

Table 2. IASI technical indicators

Spectral range	$645\ \text{cm}^{-1}$–$2760\ \text{cm}^{-1}$
Spectral resolution	$645\ \text{cm}^{-1}$–$1210\ \text{cm}^{-1}\ \Delta v \leq 0.35\ \text{cm}^{-1}$
	$1210\ \text{cm}^{-1}$–$2000\ \text{cm}^{-1}\ \Delta v \leq 0.39\ \text{cm}^{-1}$
	$2000\ \text{cm}^{-1}$–$2450\ \text{cm}^{-1}\ \Delta v \leq 0.45\ \text{cm}^{-1}$
	$2450\ \text{cm}^{-1}$–$2760\ \text{cm}^{-1}\ \Delta v \leq 0.50\ \text{cm}^{-1}$
Radiometric temperature range	4–315 K
Scanning range	96° 40′ (both sides of sub-satellite points ±48° 20′)
Number of measured points per line	30
Scan time per row	8 s
Number of fields of view	2×2 circle pixels
Each pixel field of view	11.00 mrad–14.65 mrad (related sub-satellite points 9–12 km)
Accuracy of spectral calibration	$2 \times 10^{-6}\ v$

2.3   GIFTS [6–8]

GIFTS (The Geostationary Imaging Fourier Transform Spectrometer) detects radiation mainly through two infrared bands (long-wave infrared 685–1130 cm^{-1} and medium-wave infrared 1650–2250 cm^{-1}) (Fig. 3). Because GIFTS uses 128×128 element infrared array imaging device, it can achieve 0.6 cm^{-1} hyperspectral resolution when the ground resolution is 4 km. The vertical resolution can be controlled within the range of 3–11 km (Table 3). By scanning the earth on the geosynchronous orbit, the

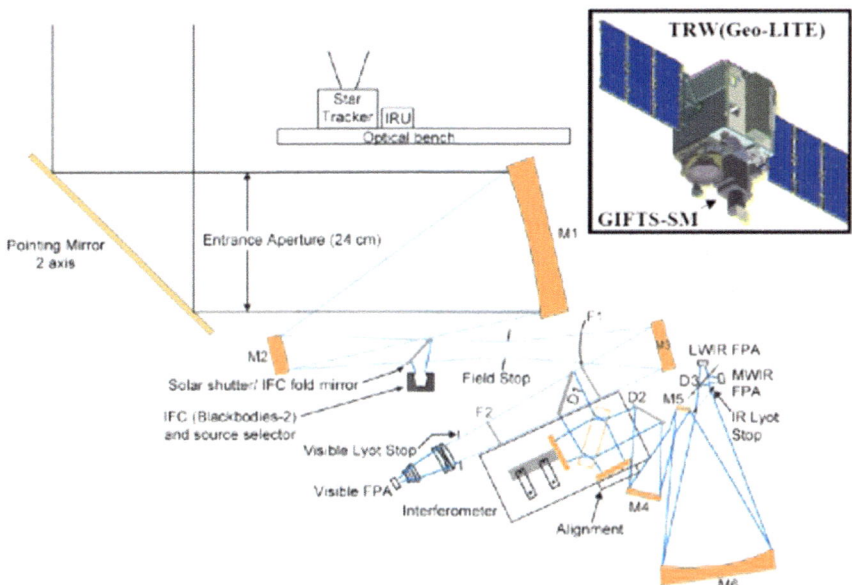

Fig. 3. Model diagram of GIFTS optical system

instrument can obtain high-resolution data of temperature, humidity, air pressure, water vapor content, wind, ozone, carbon dioxide, chemical composition and so on. Though Fourier transform calculation, these data can be effectively used in the meteorological forecast, observation, and other scientific research.

Table 3. Main technical indicators of GIFTS

Spectral range	0.4–0.8 μm 4.44–6.06 μm (2250–1650 cm^{-1}) 8.85–14.60 μm (1130–685 cm^{-1})
Spectral resolution	0.6 cm^{-1}
Spatial resolution	1 km@visible 4 km@infrared
Optical aperture	240 mm
Optical field of view	0.814° × 0.814°
Visible light detector	512 × 512 pixels
Infrared detector	128 × 128 pixels
Weight	150 kg
Power	330 W

The main features of GIFTS are as follows:

(A) Moving mirror interferometric splitting

GIFTS uses moving mirror interferometric splitting to achieve the parallel interferometric detection of the instantaneous field of view. At the same time, it is conducive to various matching modes of different spectral resolutions and detection rates. The spectral resolution of GIFTS ranges from 0.6 cm^{-1} to 36 cm^{-1}. High resolution is used in regional detection, while lower resolution is used in global exploration.

(B) Two detection bands

GIFTS selects 685–1130 cm^{-1} and 1650–2250 cm^{-1} bands, which are different from the previous detection bands. It uses the latest research results from Wisconsin University. Detection with these two bands will be more effective than the previous three bands. The main reason is that the spectral pollution of NO_2 and CH_4 can be avoided.

(C) Array detector

GIFTS uses a 128 × 128 medium-wave infrared and long-wave infrared detector array. Large-scale parallel (instantaneous array field of view) detection is realized, which makes it possible to detect meteorology in a wide range.

(D) Optical path refrigeration

GIFTS uses mechanical refrigeration to cool the detector to the 65 K working temperature to improve the performance of the detector. The interferometer system and its subsequent optical path are cooled to 150 K, the temperature of the telescope is below 220 K, and the temperature of the pointing mirror is below 290 K to reduce the interference background of the instrument.

(E) Blackbody internal calibration

GIFTS uses two blackbodies for internal calibration to determine the response and background effect of the rear part of the telescope. The influence of pre-optics is taken into account in cold space observation.

3 Development of Domestic Spaceborne Infrared Fourier Spectrometer

3.1 FY-4 [9]

The FY-4 NO. 1 star was successfully launched on December 11, 2016 at the Xichang satellite launch center. On December 17, 2016, it was fixed on the equator 99.5° east of the east longitude. The satellite carried the first interferometric infrared Fourier spectrometer, the atmospheric vertical detector, operating in the geostationary orbit.

FY-4 interferometric atmospheric vertical detector is a space imaging infrared Fourier transform spectrometer. The basic components are an interferometer, imaging system, pointing system, detector, refrigeration system, calibration system, electronics system, mechanical structure.

Infrared radiation from the ground is refracted into the telescope through the scanning mirror. The beam diameter is compressed after passing through the telescope, and then the infrared light from the ground point of the quasi-telescope is changed into parallel light through the collimator, and then reaches the beam splitter.

The beam splitter is a half translucent optical element, on which beam splits. One part of the light is reflected by the beam splitter to reach the fixed mirror and reflected back to the beam splitter by a fixed mirror. Another part of the light reaches the moving mirror through the beam splitter and reflected back to the beam splitter by the moving mirror. The beam splitter divides the rays of the two paths back to the beam splitter into two separately. Then, half of the light returns to the collimator and then to the ground along the original telescope and scanner; half of the light is directed to the splitter, which is theoretically half of the light reflected by the fixed mirror and half of the light reflected by the automatic mirror.

Since the moving mirror is constantly moving, the light path from the automatic mirror is constantly changing relative to the light path from the fixed mirror. The light passes through the color splitter and is divided into two bands of light. A long band is 700–1130 wavenumbers, i.e., 8.85–14.3 microns; a medium band is 1650–2252 wavenumbers, i.e., 4.44–6.06 μm. Then, these two bands of light are focused on the detector by subsequent optics.

The whole optical system is an imaging optical system, and the detector is conjugated to the ground. Two different detectors are used to detect the corresponding bands. Each detector has 4 rows of sensitive elements, with 32 pixels per column. The gap between columns and columns is the width of the column sensitive element.

A small step away from the scanning mirror is just the width of the detector's image on the ground moving a column so that the area between the columns can also be detected. The scanning of different areas on the ground is realized by rotating the scanning mirror in one big step and one small step in turn. When the scanning mirror points to the H standard, it will stop there. With the motion of the mirror, the intensity of light on each detector will change. The change of the intensity signal of a dynamic mirror is a signal of an interferogram. An interferogram is obtained on each pixel, so 256 interferogram signals are simultaneously obtained on 256 pixels. Then the interference signal is converted to data by the infrared signal electronics acquisition system (Fig. 4; Table 4).

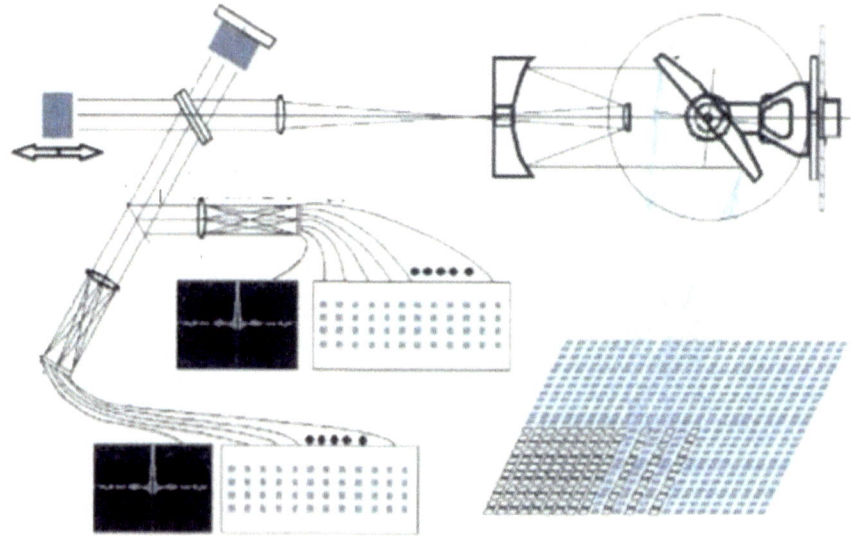

Fig. 4. Technical scheme of "vertical cloud detector" for FY-4

Table 4. Main technical indexes of FY-4 atmospheric vertical detector

Spectral range	4.44–6.06 µm (2252–1650 cm^{-1})
	8.85–14.30 µm (1130–700 cm^{-1})
Spectral resolution	0.8 cm^{-1}
	1.6 cm^{-1}
Spatial resolution	2 km@visible
	16 km@infrared
Instantaneous field angle	448 ur
Quantization level	13 bit
Infrared detector	32 × 4 pixels
1000 * 1000 detection time	35 min
5000 × 5000 detection time	67 min

After the launch of the atmospheric vertical detector, temperature control and unlocking were carried out first, and then heating and decontamination were carried out for about 45 days. Finally, the machine was turned on and cooled. When the temperature reached, the infrared channel and the interferometer were opened, and everything worked well. The following figures (Figs. 5 and 6) are an interferogram and a spectral map retrieved on the earth without a region.

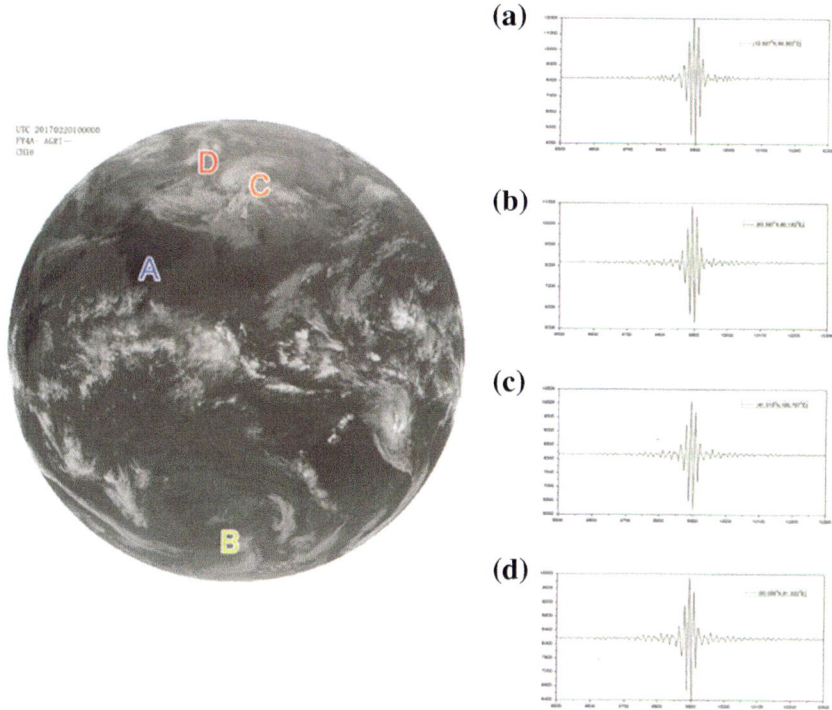

Fig. 5. Interference maps of different regions

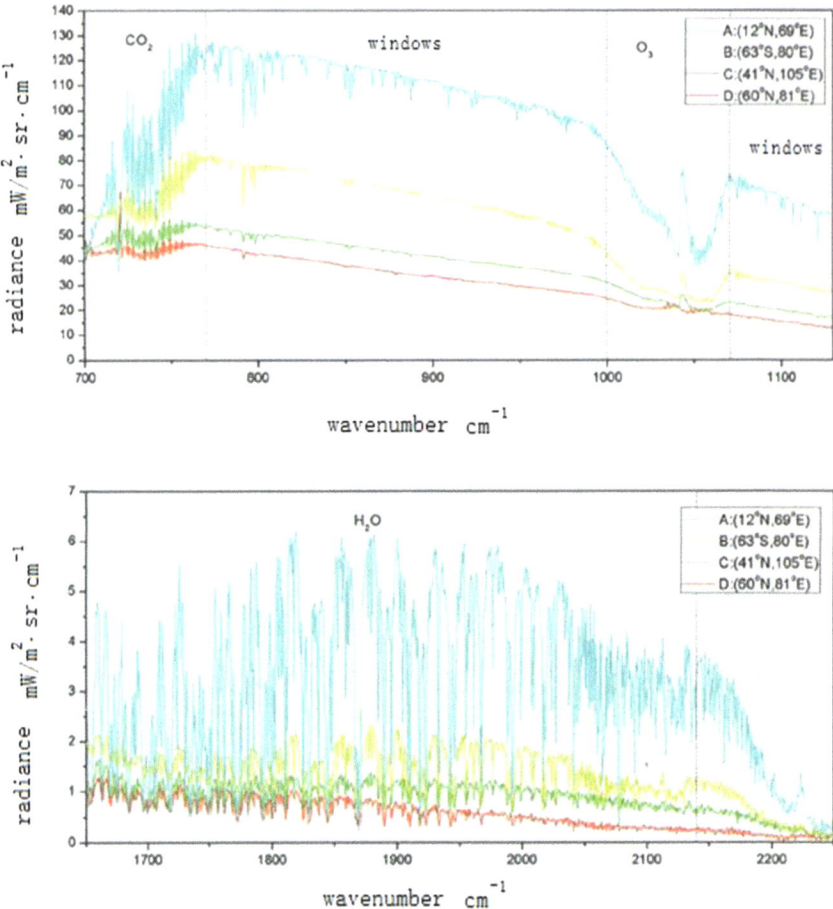

Fig. 6. Corresponding atmospheric spectra obtained by inversion

There are more than 1600 detection channels in the interferometric atmospheric vertical detector. The contribution of different altitudes to the infrared radiation of different detection channels is different. According to these differences, the three-dimensional structure of atmospheric temperature and humidity can be retrieved. Figure 7 shows the temperature distributions at different heights retrieved from the interferogram data transmitted by the instrument on orbit.

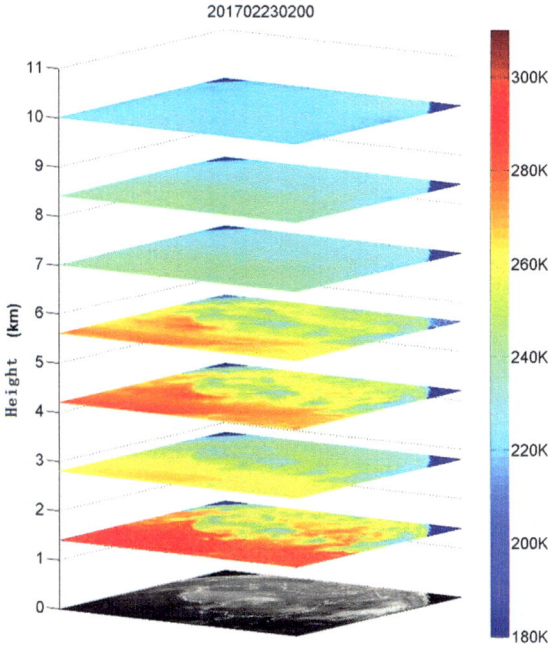

Fig. 7. Temperature distribution at different heights

3.2   FY-3 [10]

The FY-3 Star D launched successfully on November 15, 2017 at the Taiyuan satellite launch center. Satellite-borne infrared hyperspectral atmospheric detector uses moving mirror Fourier interferometric spectroscopy technology to realize infrared hyperspectral resolution detection of the atmosphere. This load makes the retrieval accuracy of atmospheric temperature and humidity profile in China more than doubled on the existing basis, and then conforms to the international advanced level.

The FY-3D Infrared Hyperspectral Atmosphere Explorer (HIRAS) will be the first hyperspectral infrared detector in the FY-3 polar orbit meteorological satellite series. It is developed by Shanghai Institute of Technical Physics, Chinese Academy of Sciences. It is an infrared hyperspectral precision optical remote sensing instrument independently developed and manufactured in China. The instrument uses two-dimensional pointing compensation, interferometer pre-position, cold optics, large cooling radiation cooling, small array infrared detector technology, showing the same capabilities as the international similar instruments on high performance, high spectral resolution, and high system sensitivity detection (Table 5).

Table 5. Main technical indicators of FY-3 infrared hyperspectral atmospheric detector

Spectral range	3.92–15.38 μm
Spectral resolution	2.5 cm^{-1}, 1.25 cm^{-1}, 0.625 cm^{-1}
Frame frequency	Four pictures per second
Number of channels	1370
Field angle	1.1°
Pixel/scan line	58
Maximum scanning angle	±50.4°
Radiometric calibration accuracy	0.7 k
Spectral calibration accuracy	7 ppm

The infrared hyperspectral atmospheric detector system is mainly composed of optical system, interferometer subsystem, and detector subsystem. As a part of collecting and analyzing earth and atmosphere radiation, the optical system is responsible for receiving infrared radiation signals from the observed targets and gathering them on the detectors, including a series of processes such as scanning, interferometric spectroscopy, convergence, color separation, and re-convergence. The telescope system is located behind the interferometer subsystem. Its main function is to collect and converge the interferometer modulation signals. The function of the relay optical system is to separate the light gathered by the telescope system according to the required band and make the light of each band enter the corresponding channel, including the turning mirror, infrared window, color splitter, field-of-view aperture, and so on. HIRAS detects and scans through a two-dimensional scanning mechanism driving a scanning mirror The scanning mirror is driven by a torque motor to do cross-rail scanning. When the scanning mirror is stationed for observation, the linear motor drives the scanning mirror to complete the track compensation movement. The main function of the interferometer is to make the input infrared radiation produce self-modulation, receive remote control instructions, output infrared interference signals, sampling control signals, and interferometer status telemetry information. The interferometer consists of two components: infrared interference and laser interference.

4 Future Development Trend of Spaceborne Infrared Fourier Spectrometer

At present, the main development direction of spaceborne infrared Fourier spectrometer technology is "one wide three high", that is, wide spectrum, high time resolution, high spatial resolution, high spectral resolution.

Improving the temporal resolution of hyperspectral sounding will be directly beneficial to short-term- and impending weather prediction, and can improve the timeliness of spectral sounding, thus enhancing the ability of meteorological observation; Widening the detection band, especially the very long band (above 14 μm), will greatly increase the ability of remote sensing information acquisition in China; improving the spatial resolution will enhance the capability of meteorological observation in small and medium scale; and further improvement of the spectral resolution will enhance the ability of fine spectral recognition.

With the advent of the era of large data, hyperspectral remote sensing data acquisition dimension is increasing, and the amount of remote sensing data acquisition is also showing an explosive growth. Efficient and effective implementation of hyperspectral remote sensing data extraction, data processing, data compression, data transmission, and data mining will be important problems to be solved in the future. Also, large data processing technology will promote the development of hyperspectral integrated data network.

5   Conclusion

As more and more satellites begin to carry infrared Fourier spectrometers, it will be a decade that infrared Fourier hyperspectral remote sensing technology develops rapidly. Many new principles, new programs, and new technologies will be implemented and applied. The instrument will develop in the direction of "one width and three heights", and large data processing technology will also promote the construction of hyperspectral Space-earth integration data network in the future.

Acknowledgements.   Funding project: No. 16XD1404100 Project; No. CXJJ-16Z243 Project.

References

1. K. Stumpf, J. Overbeck. CrIS Optical System Design [J]. SPIE, 2002, Vol. 4486
2. Ronald J. Glumb, David C. Jordan, and Joseph P. Predina. The Crosstrack Infrared Sounder (CrIS) [J]. Proceedings of SPIE, 2000, Vol. 4131
3. Dai zuoxiao, Spatial Fourier splitter detection[D], Shanghai Institute of Technical Physics, 2004(in Chinese)
4. Wang zhanhu, Study on Optical System of Space-borne Interferometric Vertical Atmosphere Detector [D], Shanghai Institute of Technical Physics, 2010 (in Chinese)
5. Liang hong, Low temperature optical problems of space interference imaging system [M], Shanghai Institute of Technical Physics, 2008 (in Chinese)
6. W. Smith, F. Harrison. The Geosynchronous Imaging Fourier Transform Spectrometer (GIFTS) [S], Official Gazette of the United States Patent & Trademark Office Patents, 2000
7. G.E. Bingham, et al. Geosynchronous Imaging Fourier Transform Spectrometer (GIFTS) Engineering Demonstration Unit (EDU) overview and performance summary [J], Proc. of SPIE, 2006, Vol. 6405, 64050F
8. W. L. Smith, et al. Geostationary Imaging Fourier Transform Spectrometer (GIFTS): Science Applications [J]. Proc. of SPIE, 2006, Vol. 6405, 64050E
9. Hua jianwen, Mao jianhua, "N0. 4 Feng Yun" Meteorological Satellite Atmospheric Vertical Detector [J], science(China), 2018, issue 1 (in Chinese)
10. Qi chengli, Gu mingjian, Hu xiuqing, Wu chunqiang. Infrared Hyperspectral Detection Technology of NO. 3 Fengyun Satellite and Its Potential Application [J]. Progress in meteorological science and technology (in Chinese)

The Ecological Function Zoning on the Hydro-fluctuation Belt in Yinzidu Reservoir

Xiaoke Zhang[1], Tongfei Feng[2], and Tao Yang[3,4(✉)]

[1] School of Public Adminstration, Hohai University, Nanjing 210098, Jiangsu, China
[2] Hebei Baoding Hydrology and Water Resources Survey Bureau, Baoding 071000, Hebei, China
[3] State Key Laboratory of Hydrology-Water Resources and Hydraulics Engineering, Hohai University, Nanjing 210098, Jiangsu, China
tao.yang@hhu.edu.cn
[4] National Cooperative Innovation Center for Water Safety& Hydro-Science, Hohai University, Nanjing 210098, Jiangsu, China

Abstract. The hydro-fluctuation belt of reservoir takes as an important ecological ecotone, its characteristics play an important role in the process of energy and material exchange in the water and land ecological system has great economic, ecological, and social value. This paper takes the Yinzidu reservoir as the research object, uses determined ecological function zoning on the basis of water level change and ecological characteristics. The main conclusions are as follows: The water level variation range of the Yinzidu reservoir always keeps between dead water level (1052 m) and normal high water level (1086 m). Its hydro-fluctuation belt area is about 8.05 km^2. And the exposed time of this place is relatively concentrated. Mulberry growth has sufficient time available. The soil has higher utilization value, suitable for planting mulberry trees. According to features of elevation changes and slope, we divided the function areas. They are water conservation forests area, ecological agriculture area, riparian buffer area, ecological restoration area, water ecological protection area, and high bedrock area. This article also develops appropriate ways to use them. The available area of hydro-fluctuation belt in Yinzidu reservoir covers an area of 2.24 km^2 provides data support for ecological restoration effect evaluation.

Keywords: Hydro-fluctuation belt · Function zoning · Yinzidu reservoir

1 Introduction

The hydro-fluctuation belt is a type of wetland, mainly distributed on the banks of rivers, lakes, and reservoirs. Reservoir hydro-fluctuation belt is different from the others in the natural conditions. It is located in the reservoir dead water level and normal high water level between the perennial affected by the change in the area of the water level, and it is the transition of land and sea ecosystem control transitional area, showing periodic dew, with water and land dual attributes and unique ecological environment

© Springer Nature Singapore Pte Ltd. 2019
L. Wang et al. (eds.), *Proceedings of the 5th China High Resolution Earth Observation Conference (CHREOC 2018)*, Lecture Notes in Electrical Engineering 552,
https://doi.org/10.1007/978-981-13-6553-9_17

characteristics. It is an area that is affected by both artificial control and natural climate [1]. As a special part of watershed ecosystem, reservoir hydro-fluctuation belt is characterized by the alternation of water and land, which makes it play an important role in the process of material and energy exchange of aquatic and terrestrial ecosystems, and has great economic, ecological and social value [2]. The research on reservoir hydro-fluctuation belt has become a hot issue in the interdisciplinary study of hydrology, ecology, and environmental science.

Ecological function zoning not only divides the ecological environment regional characteristic through the integration and the classification way, but also is significant to reveal similarities and the differences of the natural ecology region [3]. For the hydro-fluctuation belt and the bank of the mountain, according to the texture and slope of the soil, Xia Pinhua, etc., divide it into gentle slope type, bay beach type, estuary type, dish slope type, and rocky bank slope type [4]. According to the ecological characteristics of various types of hydro-fluctuation belt, Zhang Hong, etc., divide it into the reservoir tail band, the soft and stacked gentle slope type, and the hard rock steep slope type [5]. According to the different sections of the terrain, Su Weici, etc., divide it into the river bend type open terrace type, bare bedrock steep type, and unstable bank-type [6]. Considering the impact of human activities, Xie Deti divides it into the hydro-fluctuation belt which in the town, in the rural area, in the island of the reservoirs and which affected by the human activities [7]. According to the different exposed time in the flooded area, Xie Huilan, etc., divide it into perennial utilization area, seasonal utilization area and temporary utilization area [8]. Foreign studies focused on the protection of riparian zone, ecological restoration, restoration and reconstruction of vegetation, management model, etc., but failed to highlight the impact of the reservoir water level fluctuation and the characteristics of the ecological environment of the drop zone [9–13].

This paper takes the Yinzidu reservoir area as the research object, on the basis of the change of water level, the characteristics of soil factors are added, the dominant ecological function of different regional units is determined, the ecological function zoning is implemented, and the ecological restoration effect of the introduction reservoir with ecological mulberry as the basis of the "Canghaisangtian" is provided for data support, In order to set up the foundation of ecological rehabilitation and management model, the function zoning planning and control of the hydro-fluctuation belt.

2 Study Area

The project of Yinzidu hydropower station is the second stage of the cascade planning of Wujiang River, located in the downstream of Sanchahe River which is the south source of Wujiang River, the junction of Pingba County and Zhijin County in Guizhou Province, 51 km from the upstream Puding hydropower station, 43 km from the downstream Dongfeng Hydropower Station and 97 km from Guiyang City. It is in the center of Guizhou power grid. The main project is to generate electricity, the total reservoir capacity of 529 million m^3, the normal water level of 1086 m, dead level 1052 m, is an incomplete year regulation reservoir. Yinzidu reservoir is narrow, with less tributaries, which is the typical Mountain canyon type. It is a seasonal regulation of the reservoir, water is urgent while releasing, causing serious erosion on both sides of the river [14–16]. Figure 1 The Sanchahe River basin part of the primer Yinzidu hydropower station.

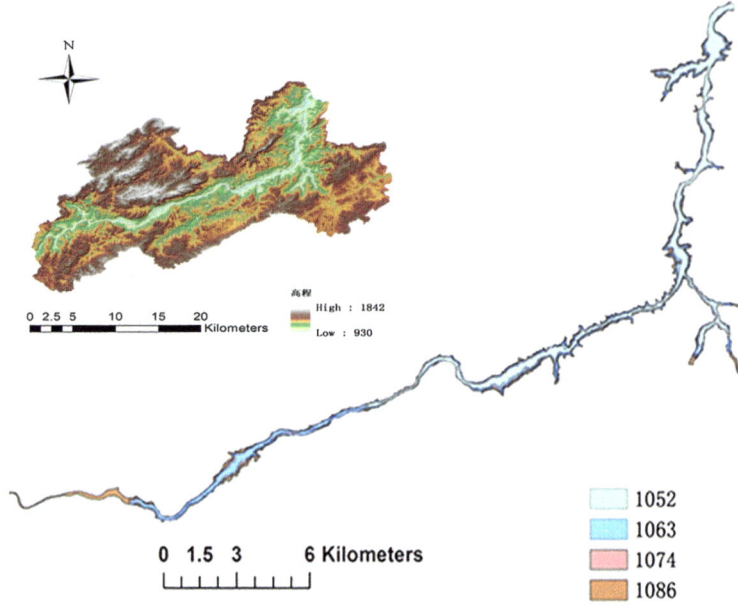

Fig. 1. The geographical location of the Yinzidu reservoir

The range of water level in the front of the Yinzidu reservoir dam is always maintained between the dead water level (1052 m) and normal high water level (1086 m). Using ArcGIS software to treat the hydro-fluctuation belt with DEM, the surface area of reservoir at Dead water level (1052 m) is about 5.8 km^2, and the corresponding surface area is 13.85 km^2 in normal high water level (1086 m), so it can be inferred that the plane area of Yinzidu reservoir is about 8.05 km^2.

The hydro-fluctuation belt can be divided into three zones according to elevation, respectively, 1052–1063 m, 1063–1074 m, 1074–1086 m; and the plane area of the corresponding several elevation belts is 2.3, 2.35, 3.4 km^2 respectively.

3   Methods

According to the dominant function principle of functional zoning, the principle of sustainable use, the principle of consistency with local natural geographical conditions, the feasibility and convenience of management and the comprehensive analysis principle [17], the zoning of the hydro-fluctuation belt of the Yizidu reservoir is carried out.

3.1 Function Zoning Indicators

(1) Elevation

The hydro-fluctuation belt in the different elevation regions corresponds to different characteristics of the water level process and the regularity of the dew-drop band. The source of a series of ecological environment changes is the change of hydrological process, which determines the dominant factor of the distribution of the vegetation in the hydro-fluctuation belt. The number of days of different elevation and the corresponding submergence period of the hydro-fluctuation belt are very important for the growth of vegetation and the development of human being, which affects the local ecological environment [18–20]. The change of water level also determines the appearance of the hydro-fluctuation belt, affects the use of different areas and the use of time, which is significant to its ecological division. According to the rule of water level fluctuation of the hydro-fluctuation belt of Yinzidu, and the characteristics of the narrow and small area of the bank, the elevation should be chosen as the index of the zoning.

According to the above analysis, the range of the hydro-fluctuation belt is divided into 1052–1057 m, 1057–1075 m, 1075–1086 m, and the area of the reservoir bank is cut out according to the ridge line around the reservoir, and based on the land use situation in the aerial film image, The bank mountain is divided into 1086–1200 m and 1200–1348 m.

(2) Slope

Slope is an important factor affecting the stability of the drop zone, which is directly related to the bank function and buffering function of the drop zone. The slope of the land in the hydro-fluctuation zone determines the size and shape of the exposed area during the land formation in the area. On the one hand, the slope is extremely related to the area, the smaller the slope, the larger the area of the land-dew in the hydro-fluctuation zone; On the other hand, with the increase of the slope within a certain range, the soil erosion of the drop zone increases, the overland velocity of the surface slope increases, and the time that flows through the drop zone becomes shorter, so the efficiency of interception of sediments and contaminants in surface runoff is significantly reduced. Because of the above two reasons, the slope is chosen as the dividing index of the hydro-fluctuation belt.

Using 25° as a boundary, >5° is considered to be steep slope, <25° considered a gentle slope. According to the investigation and theoretical analysis of the domestic hydro-fluctuation belt, the hydro-fluctuation belt of Yinzidu Reservoir is considered in the area where the slope >25°. The original farmland garden and other soil will basically disappear in a period of time (approximate years) in the rain erosion and water level fluctuation wash down, while revealing the bedrock, vegetation will not be normal survival anymore. Therefore, in the type division, >25° matrix is treated as bedrock, and the main matrix is regarded as the sediment of <25° [21, 22].

(3) Soil factor

Soil is the basic carrier of plant growth and an important part of the hydro-fluctuation belt. The physical and chemical properties of soil affect the growth of vegetation, and determine the development and stability of ecosystems. Yinzidu Reservoir construction carried out reservoir migration, the soil sampling points distribute in these immigrant villages, the average bulk density of the soil is within the range of 1.3–1.5 g/cm^3, and the total porosity of soils is basically maintained at 40–50%. According to the analysis of soil physical and chemical properties of sampling points, the soil physical properties of the region showed no obvious difference, which corresponded to the upper and lower layers of each elevation zone; but the chemical properties of soil showed great difference, especially soil organic matter. The content of organic matter in soil was between 20 and 60 g/kg, total nitrogen content was between 0.5 and 2.5 g/kg, total phosphorus content was 0.4–1.2 g/kg, total potassium content was 5–16 g/kg, providing sufficient nutritional conditions for the growth of mulberry trees. Combined with aerial imagery and land use data in the area of the Yinzidu Reservoir, we search for the suitable "Canghaisangtian" development area.

(4) Water level fluctuating

The frequency analysis of annual rainfall data of Anshun and Guiyang meteorological stations in 1996–2015 was selected, according to 25, 50, and 75%, the annual precipitation data corresponding to the abundance, flat and low-flow year were 1452.85, 1065.93, and 907.66 mm, respectively. The year of rainfall close to abundance, flat and low-flow year is selected as the corresponding represent to discuss the rainfall-water level process and the corresponding dew-time law of abundant, flat and low-flow years respectively.

Annual average water level in high flow year is 1066.98 m, the highest water level of 1085.74 m reached at the end of July, the lowest water level of 1053.29 m dropped to the beginning of June, the water level has reached a range of 32.45 m; the average level of the median water year is 1062.76 m, the highest water level 1082.34 m came at the beginning of December, in mid-April the water level reduced to the lowest of 1052.23 m, its water level range reached 30.11 m; The annual average water level in dry year is 1062.23 m, reaching the highest water level in early January of 1084.59 m, at the beginning of June the water level reduced to the lowest water level of 1052.65 m, and its water level range reached 31.94 m. There are more than 250 days when the water level lower than 1065 m, so the mulberry trees have plenty of time for growth.

3.2 Function Zoning Method

The water flow in the upstream of Yinzidu Reservoir is uncertain,and the water level uplift effect is different in size because of the change of water level in front of the dam. These reasons together led to the change of water level elevation of the reservoir along the slope is different from that of the dam, but the overall formation gap is still around 30 m (1052–1086 m).Therefore, in order to facilitate the acquisition of the range of the hydro-fluctuation belt, we used a resolution of 10 m topographic data to extract the 1052 m and 1086 m contour lines as the range of the reservoir hydro-fluctuation belt on the ArcGIS software, according to the level of the reservoir water levels. Referring to the high spatial resolution aerial data (resolution 0.2M), the ridge line of the reservoir bank mountain is identified as the range of the Bank mountain area, as the corresponding hydro-fluctuation belt and the topographic map of the reservoir bank, and the classification according to the elevation (category 5) is made. The ArcGIS software makes a slope range of the study area, which is divided into two categories: greater than or equal to 25° and less than 25°. The extraction zone and the elevation distribution zone of the reservoir and the slope partition are placed in the ArcGIS software for overlay analysis and partitioning. The results of partition are adjusted according to the aerial image and land use situation, and the result of the final functional partitioning is obtained.

4   Results

4.1   Ecological Functional Zoning of the Reservoir Bank and the Hydro-fluctuation Belts

The reservoir bank has a vital impact on the function of the hydro-fluctuation belt and the ecological environment. Therefore, some reservoir banks are also taken into account when carrying out ecological function zoning. According to the principle and method above, combined with the ecological functions of different hydro-fluctuation belts, the hydro-fluctuation belt and the reservoir bank of the Yinzidu Reservoir are divided into: Water conservation forests area, Ecological agriculture area, Riparian buffer area, Ecological restoration area, Water ecological protection area and High bedrock area.

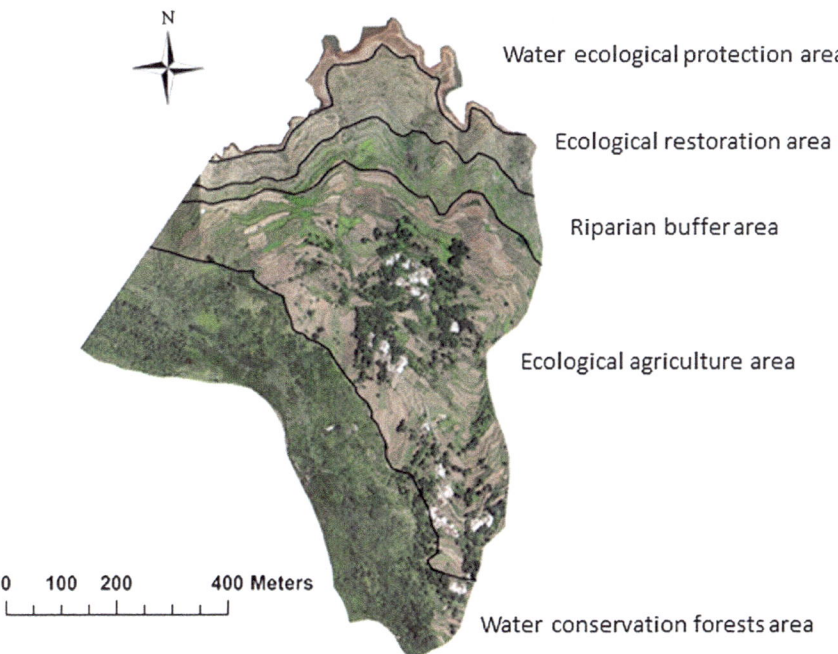

Fig. 2. The ecological function zoning on the hydro-fluctuation belt and bank mountain of the Yinzidu reservoir

In this functional division, Water conservation forests area and Ecological agriculture area are located in the reservoir bank mountain part, they are generally not flooded by the reservoir and are in a usable state for a long time; Library marina buffer area, Ecological restoration area and Water conservation area belong to the range of the hydro-fluctuation belt, and the water level periodically rises and falls, and the available time is different. According to the location of the different zones of the hydro-fluctuation belt and the bank of the reservoir, the ecological functions and utilization methods will be very different.

(1) Water conservation forest area. The area is located at a higher elevation of the reservoir bank and above the ridgeline and is suitable for forest growth. The area is located at the edge of all districts. The area of the area is about 5.53 km^2. There are some dense forests in this area, which not only protects the soil, but also reduces the scouring effect of heavy rain on the underlying surface. The role of secondary distribution reduces the direct runoff of the ground to a large extent, thereby reducing the amount of water entering the reservoir and cutting off and protecting the reservoir. The area has the role of buffering and regulating precipitation, consolidating soil and regulating climate.

(2) Ecological agriculture area. The area is located in the middle of the surrounding reservoir, below the water conservation forest area, above the normal high water level. From the situation of each ecological area in Fig. 2, the area of ecological agriculture area is the largest in several districts, reaching 32.40 km^2, which is also an important population gathering area for agricultural development zones. In the process of regional development and utilization, some pollutants may be generated. It is necessary to carry out strict environmental supervision zones and ecological industry adjustments in the zone to ensure the minimization of the impact on the water environment of the reservoir area.

(3) Riparian buffer area. The area is located above 1075 m above sea level and the area is about 2.26 km^2. Library marina buffer area is divided considering the dual needs of the sustainable use of water and soil resources and the sustainable economic and social development around the reservoir. The time that the area is exposed in a year can reach 250 days or more whether in the dry years or high level years, and it can naturally grow some water-resistant plants. The buffer zone has the function of stabilizing the river bank, regulating flood storage, and improving flood control capacity.

(4) Ecological restoration area. The area is located below Library marina buffer area, above Water conservation area, the elevation is between 1057 and 1075 m, the area is about 3.35 km^2, and the exposure time is between 120 and 250 days. Some vegetation grows seasonally in the area, and vegetation with strong vitality such as bermuda root can be grown within a few days of exposure. As the last barrier to ensure the safety of reservoir water quality, the area can intercept a large amount of sediment and non-point source pollutants brought by soil and water loss on the land bank, and reduce reservoir siltation and pollution.

(5) Water ecological protection area. The area is located above the dead water level of the reservoir, with an altitude of 1057 m or less. The area of the area is about 0.9 km^2, and the exposure time of the area is less than 120 days. Because most of

the time is flooded, there is mainly a small amount of aquatic vegetation in the area. The area is directly bordered by the reservoir, and its ecological function is to supply drinking water sources, regulate floods, regulate regional microclimates, beautify the environment and proliferate aquatic organisms, and maintain ecological balance in the area.

6) High bedrock area. The area is mainly located near the lower edge of the ecological agriculture area, with an area of about 11.88 km^2. The area is all slopes above 25°, the surface soil is basically washed away, the bedrock is exposed, the vegetation is scarce, and the species is single. When the water level rises, the area is soaked by river water, eroded by river water, and the rain is washed away, so the structure is unstable. Some engineering measures should be taken to protect the area from rock clogging (Fig. 2).

4.2   Ecological Function Zoning of Reservoir Hydro-fluctuation Belt

According to the functional zoning index, available area distributed in these immigrants village, which were Tangnai village Puding county, PuXiang village Puding county, Gezhi village PingBa county, Silahe village PingBa county, Xiangxiao village Zhijin county and Yinzidu village Zhijin county upstream to downstream in turn, each village available area is 0.22, 0.55, 0.12, 0.55, 0.34, 0.38 km^2 in order. So the available area of hydro-fluctuation belt in Yinzidu reservoir covers an area of 2.24 km^2, provides data support for ecological restoration effect evaluation (Fig. 3).

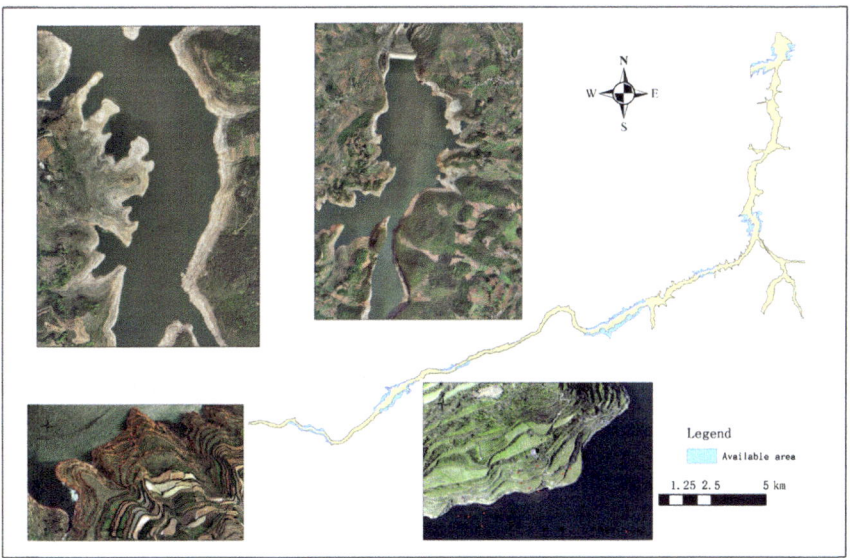

Fig. 3. The available area distribution on the hydro-fluctuation belt of the Yinzidu reservoir

5 Conclusions

Water resource is rich in our country. Much dam construction dammed the river, at the same time, it is formed a wide range of down belt. The reasonable utilization of reservoir drainage zone is urgent, and the premise is that the down belt can be scientifically planned and zoned. Based on the high spatial resolution images, the research conclusions are as follows:

(1) The water level variation range of the Yinzidu reservoir always keep between dead water level (1052 m) and normal high water level (1086 m). Its hydro-fluctuation belt area is about 8.05 km^2. And the exposed time of this place is relatively concentrated. Mulberry growth has sufficient time available.

(2) According to features of elevation changes and slope, we divided the function areas. They are water conservation forests area, ecological agriculture area, riparian buffer area, ecological restoration area, water ecological protection area and high bedrock area.

(3) The function zoning factors of hydro-fluctuation belt in Yinzidu reservoir are on elevation, slope, water level changes and soil indicator. The available area of hydro-fluctuation belt in Yinzidu reservoir covers an area of 2.24 km^2, provides data support for ecological restoration effect evaluation.

Acknowledgements. Foundation items: Supported by a grant from Ministry of Water Resources (201501032), the Fundamental Research Funds for the Central Universities (2018B21414).

References

1. Cheng Ruimei, Wang Xiaorong, Xiao Wenfa, et al. Advances in studies on water-level-fluctuation zone [J]. Scientia Silvae Sinicae, 2010, 46(4): 111–119.
2. Zheng Haijin, Yang Jie, Xie Songhua. Summary of study on fluctuation belt of reservoir Area in China [J]. Soil and Water Conservation in China, 2010, 39(06): 26–29.
3. Lei Bo, Yang Chunhua, Yang Sanming et al. GIS-based division of ecological types and their characteristics of water-level-fluctuating zone in the Three Gorges Reservoir of Yangtze river [J]. Chinese Journal of Ecology, 2012, 31(8): 2082–2090.
4. Xia Pinhua, Lin Tao, Deng Hexia, et al. The classification on the drawdown area of Hongfeng lake reservoir and its ecological restoration in Guizhou Province [J]. Soil and Water Conservation in China, 2011, 34(06): 58–60.
5. Zhang Hong, Zhu Ping. The classification of the water-level fluctuating zone in the Three Gorges Reservoir based on the interactive techniques of RS and GIS: A case study of Kaxian in Chongqing [J]. Remote Sensing for Land& Resources, 2005, 65(03): 66–69.
6. Su Weici, Yang Hua, Zhao Chunyong, et al. A preliminary study on land exploitation and utilization models of water-level-fluctuating(WLFZ) in the Three Gorges Reservoir area of Chongqing [J]. Journal of Natural Resources, 2005, 20(03): 326–332.
7. Xie Dengti, Fan Xiaohua, Wei Chaofu. Effects of the riparian zone of the Three-Gorges Reservoir on the water-soil environment of the reservoir area [J]. Journal of Southwest University (Natural Science), 2007, 35(01): 39–47.

8. Xie Huilan, Zhang Xueyong. The rational utilization of land resources in Huangbizhuang resevoir drawdown area [J]. Resources Development and Conservation, 1991, 35(02): 96–98.
9. Azza N, Denny P, VAN DE Koppel J, et al. Floating mats: their occurrence and influence on shoreline distribution of emergent vegetation [J]. Freshwater Biology, 2006, 51(7): 1286–1297.
10. Whigham, D.F., Ecological issues related to wetland preservation, restoration, creation and assessment [J]. Science of the Total Environment, 1999. 240(1–3): 31–40.
11. Anbumozhi, V., J. Radhakrishnan and E. Yamaji, Impact of riparian buffer zones on water quality and associated management considerations [J]. Ecological Engineering, 2005. 24(5): 517-523.
12. Nakamura, F. and H. Yamada, Effects of pasture development on the ecological functions of riparian forests in Hokkaido in northern Japan [J]. Ecological Engineering, 2005. 24(5): 539–550.
13. Holmes, P.M., et al., Guidelines for improved management of riparian zones invaded by alien plants in South Africa [J]. South African Journal of Botany, 2008. 74(3): 538–552.
14. Zhai Gang. The determination of the flood flow during construction period in Yinzidu resevoir [J]. Guizhou Water Power, 2002, 16(01): 64–66.
15. Wang Shilong. On engineering geological problems in Yinzidu hydropower station [J]. Dam and Safety, 2004(02): 8–10.
16. Wang Shilong. The understanding of the cause of the induced earthquake in the Yinzidu hydropower station [J]. Guizhou Water Power, 2003(05): 33–35.
17. Yin Wei, Lei Junshan, Ye Min, et al. The wetland ecological function zoning and protection countermeasures in Danjiangkou reservoir [J]. Yangtze River, 2008, 39(23): 80–82.
18. Wang Yechun, Lei Bo, Yang Sanming, et al. Concentrations and Pollution Assessment of Soil Heavy Metals at Different Water-level Altitudes in the Draw-down Areas of the Three Gorges Reservoir [J]. Environmental Science, 2012(02): 612–617.
19. Wang Yechun, Lei Bo, Zhang Sheng. Differences in vegetation and soil characteristics at different water-level altitudes in the drawdown areas of Three Gorges Reservoir area [J]. Journal of Lake Sciences, 2012(02): 206–212.
20. Sun Rong, Liu Hong, Ding Jiajia, et al. Quantitative Analyses of Plant Communities in the Hydro-fluctuation Area in Kaixian County After Impoundment of the Three-Gorge Reservoir [J]. Journal of Ecology and Rural Environment, 2011(01): 23–28.
21. Zhang Baolei, Zhang Shumin, Zhou Qigang, et al. Dynamics in land use and ecosystem service – a case of Daning river watershed in Three-Gorge Reservoir area[J]. Resources and Environment in the Yangtze Basin, 2007(02): 181–185.
22. Zhong Ronghua, Cheng Xupeng, Huang Jiangcheng, et al. Spatial distribution of water level fluctuating zone in the Xiaowan Reservoir, China [J]. Journal of Yunnan University(Natural Sciences), 2017,39(6): 1104–1110.

High-Speed Object Tracking with Dynamic Vision Sensor

Jinjian Wu$^{(\boxtimes)}$, Ke Zhang, Yuxin Zhang, Xuemei Xie,
and Guangming Shi

School of Artificial Intelligence, Xidian University, Xi'an 710071, Shaanxi,
China
jinjian.wu@mail.xidian.edu.cn

Abstract. High-speed object tracking is still a great challenge for video processing. Traditional cameras can hardly capture the motion trajectory of the high-speed moving object. With differential logarithmic photodetector and nanosecond response latency to fast stimuli, dynamic vision sensor (DVS) is extremely sensitive to the moving object (especially for the object with high speed). However, existing object tracking algorithms, which are limited by their frame-by-frame processing mode, are no longer suitable for DVS. In this work, we introduce a novel event coherence detection algorithm for high-speed objective tracking. The moving target is determined by judging the coherence of the event according to the event distribution at a certain moment. Experimental results demonstrate that the proposed algorithm can accurately track the small objects with high speed. Meanwhile, the proposed algorithm performs efficiently, which can run in real time.

Keywords: Object tracking · Dynamic vision sensor · Event-based vision · High-speed object

1 Introduction

High-speed object tracking is useful in many occasions, e.g., analyzing the motion trajectory of tennis and golf (which improves the skills of athletes), tracking unexpected invading objects in specific space, and so on. Conventional frame-based cameras, for which the frame rate is typically on the order of 30 fps, will always miss some useful dynamic information. As a result, such kinds of cameras are limited to be applied in tracking high-speed moving objects. Yet advances in digital camera technology toward higher frame rates have yielded the ability to better capture high-speed motion, but at cost of expensive storage space and bandwidth. Meanwhile, highly computational power is also demanded for the information processing on all pixels (including these pixels which is the absence of motion information in the scene).

Event-based computer vision, which addresses event representation (AER), provides a sound solution to high-speed vision problems [1]. This newly developed

© Springer Nature Singapore Pte Ltd. 2019
L. Wang et al. (eds.), *Proceedings of the 5th China High Resolution Earth Observation Conference (CHREOC 2018)*, Lecture Notes in Electrical Engineering 552,
https://doi.org/10.1007/978-981-13-6553-9_18

discipline is motivated by mimicking biological visual systems [2]. The dynamic vision sensor (DVS) reacts to changes in contrast, which are then converted into a stream of asynchronous time-stamped events [3]. The reduction of redundant information makes this technique promising for high-speed tracking.

The use of event-based vision requires the development of time-oriented event-based algorithms, in order to benefit fully from the properties of this new framework [4]. Feature detection and tracking methods for frame-based cameras are well established. For example, correlation filtering has gained attention in single object tracking due to computational efficiency [5–7]. Convolutional Neural Network (CNN) is also a popular solution in visual tracking field by exploiting the semantic features extracted through layers [8–10]. For multiple objects, tracking by detection has become the leading paradigm [11–13]. Although these algorithms have achieved good results on object tracking problems, they are frame-based and no longer suitable for DVS. The reason is as follow, DVS produces a discrete three-dimensional event stream similar to a flowing point cloud and intrinsically it has no concept of frame. If we choose representation by compressing the event stream for a fixed temporal interval into one frame, it will be affected by the velocity, because the size of a temporal window is dependent on the speed of the object in the environment. Under- or over-estimating the temporal interval will result in motion blur or incomplete representation [13]. So it is only possible to filter out irrelevant events and preserve coherent events from the perspective of events. Furthermore, visual tracking is a natural application for DVS, because it has the characteristics of only responding to changes. A moving target generates a continuous stream of events in which each pixel along the trajectory triggers an event, resulting in a detailed description of the target's path even for high-speed motions. In other words, the full path is always supplied by DVS [14].

Recently, several methods for event camera tracking have been presented. In [13, 19], Harris corners are detected using the event density over a temporal window. This work focuses on feature tracking in the event stream, but every single event has to be calculated so that it cannot be applied for real-time object tracking. [14] introduces a particle filter which leads to a loss in different velocities between camera and target moving. However, its proposed algorithm required a priori knowledge to determine the objects to track. [15] presents an event stream representation by approximating the 3D geometry of the event stream with a parametric model. Whereas, the targets in the training data have the same motion trajectory and velocity, therefore it cannot be used for any scene. The method of [16–18] uses both the events and frames to detect and track object. They detect features from grayscale frames then track on the event stream. However, this method needs to establish correspondences between the events and the frames. In summary, all of these existing methods need to accumulate a time window to create an information representation.

In this paper, we propose a novel event coherence detection algorithm, which uses clustering to analyze whether there is a correlation between events. The algorithm rejects the effectiveness judgment based on a single event in the above algorithm. It presents a new idea that the moving target is determined by judging the coherence of the event according to the event distribution at a certain moment (see Sect. 2.1 for details). Our algorithm has the following contributions compared to the algorithm described above, on the one hand, the algorithm directly calculates on the 3D events, which maintains the originality of the DVS output signal and abandons the concept of the frame. On the other hand, a number of events processed by the algorithm at the same time is much larger than 1, so that it can run in the real-time object tracking. Finally, because the inspiration is based on scanning the event in a row array of DVS, which characteristic makes the algorithm universal for all data.

2 Algorithms

An event camera responds to the variation of the light level in environment [14], the data from such a camera consists of an asynchronous stream of events in visual-temporal space. To be more specific, a moving target generates a continuous event stream in which each pixel along trajectory triggers an event. Hence we can take advantage of the dense spatiotemporal stream in event space. Once the trajectory is detected, all we require is to maintain tracking.

2.1 Event Coherence Detection Algorithm

As mentioned before, compressing a period of event stream into a frame is not feasible. In this session, we will present our novel algorithm which is designed from the ground up to most efficiently make use of event camera's advantages. As is known to us all, in order to distinguish effective events and noisy points, it is crucial to preserve the coherence among different events. The reason why we design this algorithm is that DVS is born with the characteristic of recording the event in response to the environmental change, in this way it scans the event in a row array and stores in binary mode. When a moving target appears in the scene, the DVS will generate multiple responses to the target at the same time. According to this characteristic of DVS, we propose our algorithm as follows:

(1) The event stream is divided into equally sized stream blocks in three dimensions. There are two reasons for this operation, on the one hand, as a single event does not provide enough information alone, we need to create an asynchronous local coherence condition for each event. On the other hand, we need to ensure that the most responsive events at the same time belong to the same object. According to the simplicity and complexity of the scene, we generally take the stream block size to 32 or 64 blocks each row and column. As is shown in Eq. (1), the *width* and *height* are temporal resolution of DVS.

$$\left\{ e_{t_1,x_1,y_1}, e_{t_2,x_2,y_2} \cdots e_{t_\zeta,x_\zeta,y_\zeta} \right\}$$
$$= \left\{ E_{t,x,y} | x = blocks : blocks : width, y = blocks : blocks : height \right\} \quad (1)$$
$$\zeta = blocks * blocks$$

(2) Here we define the center point of the event coherence as K, which has an associated timestamp, t, pixel location, $\langle x, y \rangle$. If the total number of events in a stream block is very small, then remove this block. As is presented in Eq. (2), for existing blocks we calculate the moment t when most events are generated. And

μ^* denotes the number at t, we call this moment the timestamp of K, the corresponding position median is taken as the pixel location of K.

$$K_t = max(count(e_i)|t)$$
$$\mu^* = count\left(e_{t,x_i,y_i}\right) \tag{2}$$
$$K_{x,y} = median\left(e_{t,x_1,y_1}, e_{t,x_2,y_2}, \ldots e_{t,x_{\mu^*},y_{\mu^*}}\right)$$

(3) Coherent event cluster C is

$$O : C = \left\{ (e_i)|\, \|e_{t_i} - K_t\| \| < = T_{thr} \,\&\, \left\| \|e_{x_i,y_i} - K_{x,y}\right\|^2 < = L_{thr} \right\} \tag{3}$$

here, taking K as the space-time center, the event e_i within the time threshold and the spatial threshold are the detected effective coherent events, which consist of the target to be tracked. The proposed event coherence detection algorithm has the advantages of faster speed, higher accuracy, and better robustness compared with the existing event-based tracking algorithm.

2.2 Single Target Tracking for DVS Events

The overall procedure of our single target tracking algorithm is presented in Listing 1. The input is the event stream with an event length of X, we denote these four parameters, x, y, ts, and polarity. First, the detectEventCoherence function applies Eqs. (1–3) throughout the space to detect coherent events. The moving target consists of these coherent events which can be calculated from Eq. (4). Then from (5) we develop a region of interest $Roi_{x,y}$ for the target of the output $Obj_{x,y}^t$. The event stream entered later is only detected in $Roi_{x,y}$.

$$\left\{ Obj_{x,y}^t \right\} = \bigcup \{(e_i)\} \tag{4}$$

$$row\_min = max\left(1, \min_x \left(\left\{ Obj_{x,y}^t \right\}\right) - L\right)$$
$$row\_max = min\left(width, \max_x \left(\left\{ Obj_{x,y}^t \right\}\right) + L\right)$$
$$col\_min = max\left(1, \min_y \left(\left\{ Obj_{x,y}^t \right\}\right) - L\right) \tag{5}$$
$$col\_max = min\left(height, \max_y \left(\left\{ Obj_{x,y}^t \right\}\right) + L\right)$$
$$Roi_{x,y} = [row\_min, row\_max, col\_min, col\_max]$$

$$\lambda = \frac{\sum\limits_{x^i,y^i} \left[Obj_{x^i,y^i}^t \cdot M \right]}{\sqrt{\sum\limits_{x^i,y^i} \left(Obj_{x^i,y^i}^t \right)^2 \cdot \sqrt{M^2}}} \tag{6}$$

If $Obj_{x,y}^t$ is empty in a detection, we will readjust the region of interest to the entire space and match the detected target to the existing model by Eq. (6). Here λ indicates the matching rate. Repeat the above steps as long as there is an event input until end of the stream.

Listing 1: Event-based single target tracking procedure

Input: Three dimension event stream with Event.x, Event.y, Event.ts, Event.pol.

Output: Detected target states $Obj_{x,y}^{t\,*}$.

---

1 : Initialize parameters;

2 : $\{Obj_{x,y}^t\} \leftarrow$ detectEventCoherence(Event, $R_{width, height}$);

3: Define model M;

4: **while** (Event) **do**

5 :   **repeat**

6 :   **if** ($\{Obj_{x,y}^t\}$!=null) **then**

7 : $Roi_{x,y} \leftarrow$ regionOf Interest($\{Obj_{x,y}^t\}$);

8:             $\{Obj_{x,y}^t\} \leftarrow$ detectEventCoherence(Event, $Roi_{x,y}$);

9:             $Obj_{x,y}^{t\,*} \leftarrow$ matchObject(M, $Obj_{x,y}^t$);

10:            $M \leftarrow$ updateModel(M, $Obj_{x,y}^{t\,*}$);

11:        **else if** ($\{Obj_{x,y}^t\}$ ==null) **then**

12:            $\{Obj_{x,y}^t\} \leftarrow$ detectEventCoherence(Event, $R_{width, height}$);

12:            Go to step 4;

13:        **end if;**

14:    **Until** end of stream;

15: **end while.**

2.3   Multiple Target Tracking for DVS Events

We present the main steps of our multiple target tracking algorithms in Listing 2. We employ a tracking method based on the detection. To be specific, we need to set detections and trackers to cooperate with each other during the tracking process. For a given stream block with a fixed number of events, apply the event coherence detection algorithm to the whole search scope as Formula (7).

$$\{Obj_{x,y}^t{}^1, Obj_{x,y}^t{}^2 ... Obj_{x,y}^t{}^k\} = \bigcup \{(e_i) \big| \| e_i - e_j \| <= L_{thr}, j = (i+1) : length(e_i)\} \qquad (7)$$

Here k represents the number of targets. Base on those events we can establish detections and preserve the targets to be detected. Then we store all the targets detected for the first time into trackers and assign specific IDs to them. For the following event stream, the detections will try to match the targets in the trackers and upgrade them with the same ID. For those targets that cannot be matched in the detections, add them to the trackers and assign them with new IDs. If there is a target in the tracker which could not match the objects in the detections after a long period of time, delete the targets' ID information.

Listing 2: Event-based multiple target tracking procedure

Input   : Three dimension event stream with Event.x, Event.y, Event.ts, Event.pol.

Output : Detected target states $\left\{Obj_{x,y}^{t\,1}, Obj_{x,y}^{t\,2} \cdots Obj_{x,y}^{t\,k}\right\}$.

1:   Initialize parameters;

2:   $\left\{Obj_{x,y}^{t\,1}, Obj_{x,y}^{t\,2} \cdots Obj_{x,y}^{t\,k}\right\} \leftarrow$ detectEventCoherence(Event, $R_{width,height}$);

3:   Define tracker T, Initialize ID_s;

4:   **while** (Event) **do**

5:        **repeat**

6:             $\left\{Obj_{x,y}^{t\,1}, Obj_{x,y}^{t\,2} \cdots Obj_{x,y}^{t\,k}\right\} \leftarrow$ detectEventCoherence (Event, $R_{width,height}$);

7:             Assignments, UnassignedTracks, UnassignedDetection $\leftarrow$
             associateDetectionsToTrackers(T, $\left\{Obj_{x,y}^{t\,1}, Obj_{x,y}^{t\,2} \cdots Obj_{x,y}^{t\,k}\right\}$);

8:             updateAssignedTracks(Assignments, T, ID_s);

9:             updateUnassignedTracks(UnassignedTracks, T);

10:           deleteLostTracks(T, ID_s);

11:           createNewTracks(UnassignedDetections, T, ID_s);

12:      **Until** end of stream;

13: **end while**.

---

3   Experiments

3.1   Dataset Collection

Two datasets were collected under two different scenes, both utilized CeleX-IV Sensor equipped with a 16 mm lens. Scenario description and experimental setup for two datasets are stated in Sects. 3.1.1 and 3.1.2.

CeleX-IV is a high-resolution high-speed sensor, with 768×640 pixels and maximum 200 Meps output. Different from the DAVIS, which is commonly used in DVS-related research, the CeleX-IV is more manipulative on parameters setting. Other than adjusting focal distance and aperture, the threshold and time resolution that determine how the pixels are triggered could also be adjusted.

Two distinctions of output events format between DAVIS and CeleX-IV should be alerted. The first one lies in the events output method, which is CeleX-IV outputs the events of a single line at each triggered timestamp. The second difference is that DAVIS keeps record of polarity information in stream of events while CeleX-IV recorded the intensity level evaluated with 9 bits.

3.1.1 Single Tennis Dataset

Single tennis dataset was collected on the playground outdoors in the natural daylight, with pedestrian moving and complicated texture in the background. The tracking target is one single tennis played in front of the camera, its motion trajectory is covered in camera's field of vision.

To compare the capturing capability and tracking effects between event-based and frame-based vision, both DVS and DV are used to record this scene. DVS and DV are attached to tripods and are placed closely, with orientation adjusted consistent towards the single flying tennis. DV and DVS are kept still during the shooting process. No extra setting was made to DV. For DVS, other than manually adjusting focus, the threshold and time resolution remains 25 and 40 ns as default.

3.1.2 Multiple Tennis Dataset

Multiple tennis dataset was collected in the indoor stadium in normal daylight without artificial lights disturbing. The tracking targets are multiple tennis thrown across the picture in the opposite direction by two people standing outside the view of camera.

Only DVS is used for recording in this dataset. DVS was kept still during the whole shooting process. Aperture and focus were adjusted accordingly to guarantee the imaging quality. Threshold was set as 60 to reduce the amount of events and time resolution remains 40 ns.

3.2 Single Object Tracking Comparison

In this section, we apply single tennis dataset to track and show comparing results on DVS and DV in Figs. 1 and 2. There are nine frames to describe the tracking results using the frame-based tracking algorithm. We can clearly see that DV-based tracking is a failure because the speed of moving target is quite fast, leading to complete loss of target. In contrast, tracking based on DVS output events can fully describe the motion trajectory of the target. As shown in Fig. 2, red scatter indicates the trajectory of the target motion, and gray scatter indicates the events output by the DVS.

Fig. 1. Results for frame-based tracking algorithm for high speed tennis, indicating bounding box unable to track the trajectory of the target

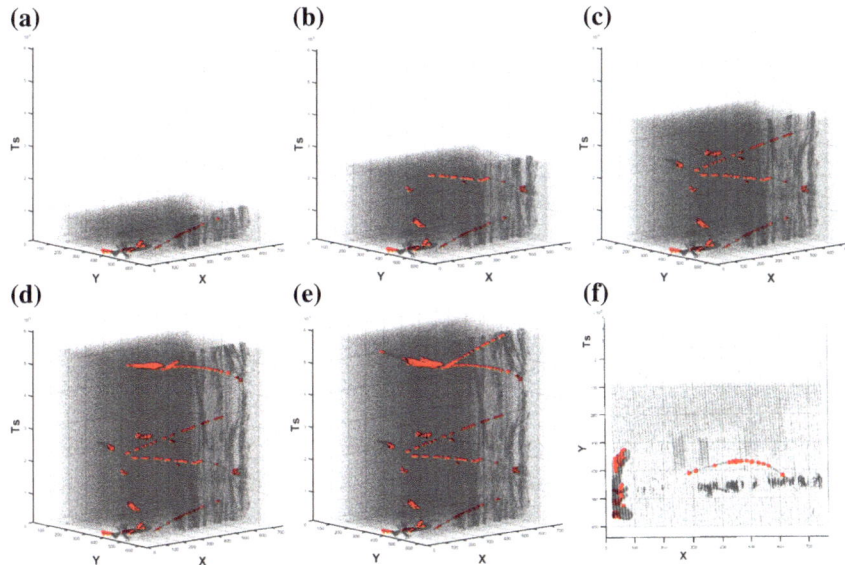

Fig. 2. Results of event-based tracking algorithm for high speed tennis. **a–e** Tracking process in three dimensions(x, y, Ts); **f** 2D plane projection of the target's spatial trajectory

3.3 Event-Based Multiple Objects Tracking

On account of our algorithm finishes the task of detection, it not only performs well in single target tracking but also multi-target tracking. We report results in Fig. 3 which can completely present the motion state of different targets. Unlike single target tracking, we need to maintain the IDs for kinds of moving targets in multi-target tracking. (f) accumulates the trajectory of the entire motion cycle of (a–e).

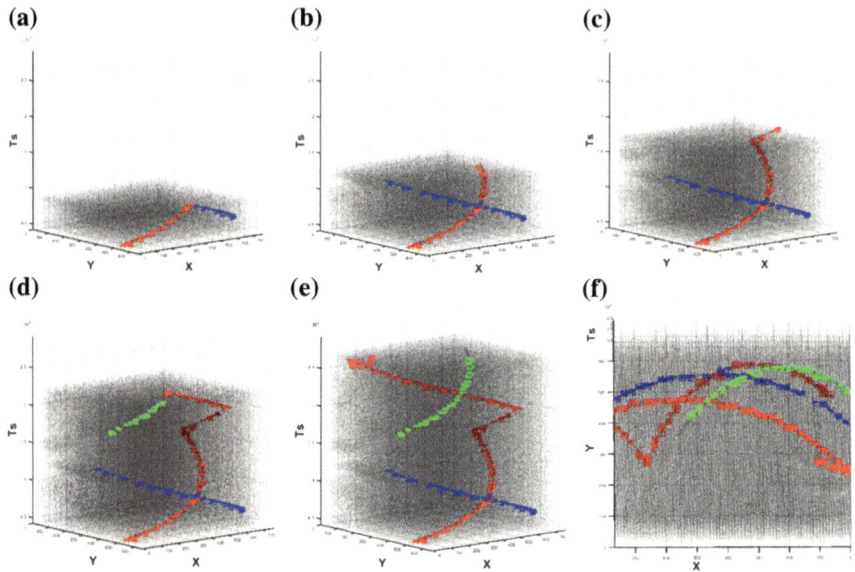

Fig. 3. Results of event-based multiple high-speed objects tracking. **a–e** 3D tracking trajectory; **f** 2D plane projection of inter-plane motion trajectory. Among all, different colors indicate different moving targets

3.4 Evaluation

As shown in Table 1, compared to V. Vasco's event-based Harris corner detection algorithm at in [14] and [20], our cluster-based event coherence detection algorithm has obvious advantages in performance, speed, and application. First, V. Vasco applies traditional image processing algorithms for the analysis of the events, while our algorithm analyzes and processes 3D event streams based on the nature of the event. As shown in Fig. 4, our algorithm is more accurate in detecting effective events than V. Vasco' algorithms because in our method the targets have the maximum number of response events at the same time. Finally, for the comparison of algorithm speed, our method runs with 10,000 events per 0.07 s contrasting to 10,000 events per 0.9 s of V. Vasco's algorithm, which addresses the high efficiency of our algorithm.

(a)   **(b)**   **(c)**

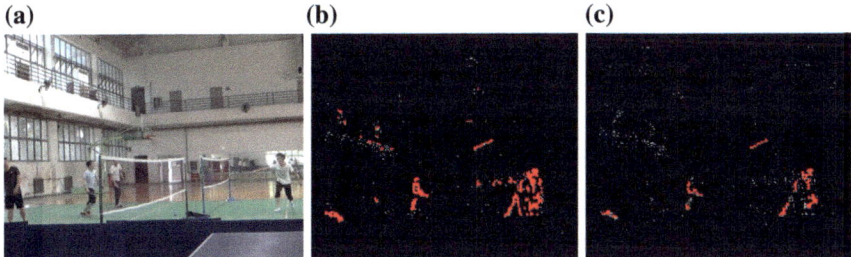

Fig. 4. **a** is an indoor badminton scene, because the window has light changes, DVS will have an event response at the window. But these events are ineffective for us. In (**c**), our algorithm only detects moving target while (**b**) algorithm based on single events also detects the window

Table 1. Comparison with existing algorithms

V. Vasco's event-based Harris corner detection	Event coherence detection algorithm
Apply traditional image processing algorithms	Event-based analysis and processing algorithm
The effective events detected don't all belong to moving targets (see Fig. 4)	All effective events detected belong to moving targets.
Slow, 10,000 events/0.9 s	Fast, 10,000 events/0.07 s

4   Conclusion

In this work, we have introduced a novel event coherence detection algorithm for high-speed object tracking. DVS produces a discrete three-dimensional event stream, which is similar to a flowing point cloud and intrinsically has no concept of frame. And thus, existing object tracking algorithms are no longer suitable for DVS. In order to track the small object with high speed, we take the advantages of unique characteristics of DVS and apply cluster-based event coherence detection algorithm on 3D event stream. By calculating the moment when the event is most generated, the center point has been set. Then coherent events have been preserved with both the time and the spatial thresholds, which also help to filter out irrelevant events at the same time. Experimental results have demonstrated that the proposed algorithm can accurately track the high-speed small objects in real-time.

References

1. T. Delbruck, "Frame-free dynamic digital vision," in Proc. Int. Symp. Secure-Life Electron., Adv. Electron. Quality Life Soc., 2008, pp. 21–26.
2. T. Delbruck, B. Linares-Barranco, E. Culurciello, and C. Posch, "Activitydriven, event-based vision sensors," in Proc. IEEE Int. Symp. Circuits Syst., May 30–Jun. 2, 2010, pp. 2426–2429.

3. P. Lichtsteiner, C. Posch, and T. Delbruck, "A 128×128 120 dB 15 μs latency asynchronous temporal contrast vision sensor," IEEE J. SolidState Circuits, vol. 43, no. 2, pp. 566–576, Feb. 2008.
4. Z. Ni, C. Pacoret, R. Benosman, S. Ieng, and S. Regnier, "Asynchronous ´ event based high speed vision for micro-particles tracking," J. Microscopy, vol. 245, pp. 236–244, 2012.
5. J. F. Henriques, R. Caseiro, P. Martins, and J. Batista. Highspeed tracking with kernelized correlation filters. TPAMI, 37(3): 583–596, 2015. 1, 6, 8.
6. M. Danelljan, G. H¨ager, F. Shahbaz Khan, and M. Felsberg. Accurate scale estimation for robust visual tracking. In BMVC, 2014. 1, 8.
7. M. Danelljan, A. Robinson, F. Shahbaz Khan, and M. Felsberg. Beyond correlation filters: Learning continuous convolution operators for visual tracking. In ECCV, 2016. 1, 2, 3, 4, 7, 8, 10, 11.
8. H. Nam, M. Baek, and B. Han. Modeling and propagating cnns in a tree structure for visual tracking. CoRR, abs/1608.07242, 2016. 7, 8.
9. L. Bertinetto, J. Valmadre, J. F. Henriques, A. Vedaldi, and P. H. Torr. Fully-convolutional siamese networks for object tracking. In ECCV workshop, 2016. 2.
10. M. Danelljan, G. H¨ager, F. Shahbaz Khan, and M. Felsberg. Convolutional features for correlation filter based visual tracking. In ICCV Workshop, 2015. 1, 8.
11. A. Bewley, G. Zongyuan, F. Ramos, and B. Upcroft "Simple online and realtime tracking," in ICIP, 2016, pp. 3464–3468.
12. N. Wojke, A. Bewley, D. Paulus, Simple online and realtime tracking with a deep association metric, CoRR, abs/1703.07402, 2017.
13. A. Milan, S. H. Rezatofighi, A. Dick, I. Reid, and K. Schindler. Online multi-target tracking using recurrent neural networks. In AAAI, 2016. 1, 2, 8.
14. Valentina Vasco, Arren Glover, and Chiara Bartolozzi. Fast event-based Harris corner detection exploiting the advantages of event-driven cameras. In IEEE/RSJ Int. Conf. Intell. Robot. Syst. (IROS), 2016. https://doi.org/10.1109/iros.2016.7759610.
15. Arren Glover, Chiara Bartolozzi. Robudt visual tracking eith a freely-moving event camera. In IEEE/RSJ Int. Conf. Intell. Robot. Syst. (IROS), 2017.
16. Mitrokhin, A., Fermuller, C., Parameshwara, C., Aloimonos, Y.: Event-based Moving Object Detection and Tracking. arXiv preprint arXiv:1803.04523 (2018).
17. Kueng, B., Mueggler, E., Gallego, G., Scaramuzza, D.: Low-latency visual odometry using event-based feature tracks. In: IEEE/RSJ Int. Conf. Intell. Robot. Syst. (IROS), Daejeon, Korea (October 2016) 16–23.
18. Zhu, A.Z., Atanasov, N., Daniilidis, K.: Event-based feature tracking with probabilistic data association. In: IEEE Int. Conf. Robot. Autom. (ICRA). (2017) 4465–4470.
19. H. Liu, D. P. Moeys, G. Das, D. Neil, S.-C. Liu, and T. Delbruck, "Combined frame- and event-based detection and tracking," in Int. Conf. on Circuits and Systems (ISCAS), 2016.
20. V. Vasco, A. Glover, E. Mueggler, D. Scaramuzza, L. Natale, and C. Bartolozzi, "Independent motion detection with event-driven cameras," in Advanced Robotics (ICAR), 2017 18th International Conference on. IEEE, 2017, pp. 530–536.

Analysis of the Algebraic Tails
of GF-3 SAR Images

Zengguo Sun[(⊠)] and Yunjing Song

School of Computer Science, Shaanxi Normal University, Xi'an 710119,
Shaanxi, China
duffer2000@163.com

Abstract. GF-3 is the synthetic aperture radar (SAR) satellite with the largest number of imaging modes in the world, which operates in 12 imaging modes. Owing to the 1-m resolution imaging mode, GF-3 satellite has become the highest resolution satellite system in the world for C-band multipolar SAR satellites. The high-resolution SAR images taken by the GF-3 satellite are rich in information. However, due to the coherent imaging system, GF-3 SAR images are affected by speckle, which degrades the quality of images seriously and makes the postprocessing of images such as edge detection, image segmentation, and target recognition extremely difficult. In this paper, we analyzed the characteristics of GF-3 SAR images from two aspects of samples and distributions. It can be seen that the GF-3 SAR images have algebraic tails. Furthermore, the above characteristics of GF-3 SAR images are verified by related experiments including the heavy-tailed Rayleigh distribution-based modeling and the maximum a posteriori (MAP) filter using the heavy-tailed prior distribution. Compared with the traditional models, GF-3 SAR images are well modeled by the heavy-tailed Rayleigh distribution that owns clear characteristics of sharp peak and heavy tail. This means that the GF-3 SAR images are statistically sharp-peaked and heavy-tailed. Additionally, the MAP filter with the heavy-tailed prior distribution achieves better performance in comparison with the traditional filters such as Lee filter and Gamma MAP filter, which is attributed to the choice of heavy-tailed prior distribution with the sharp peak and heavy tail. Therefore, based on the experiments above, we verify that the GF-3 SAR images own sharp-peaked and heavy-tailed statistical characteristics.

Keywords: GF-3 SAR images · Image modeling · MAP filter · Heavy-tailed distributions

1 Introduction

The GF-3 satellite, which has been launched on August 10, 2016 by China Academy of Space Technology, is the first Chinese civil C-band synthetic aperture radar (SAR) and has 12 imaging modes with a fine spatial resolution up to 1 m. GF-3 satellite is not subject to weather conditions such as cloud and rain, and it serves the oceans, disaster reduction, water conservancy, meteorology, and many other fields. Nevertheless, due to the coherent imaging system, speckle appears in GF-3 SAR images. One major challenge of the SAR image processing is the impact of speckle. In order to obtain high-

© Springer Nature Singapore Pte Ltd. 2019
L. Wang et al. (eds.), *Proceedings of the 5th China High Resolution Earth Observation Conference (CHREOC 2018)*, Lecture Notes in Electrical Engineering 552,
https://doi.org/10.1007/978-981-13-6553-9_19

quality despeckling result, we have to accurately describe the statistical characteristics of GF-3 SAR images [1, 2]. Therefore, it is very necessary to study the statistical characteristics of GF-3 SAR images. In this paper, we analyze the characteristics of the GF-3 SAR images from two aspects. Analysis of sample distribution and statistics demonstrate that GF-3 SAR images have sharp-peaked and heavy-tailed, respectively. Thus GF-3 SAR images have sharp-peaked and heavy-tailed statistical properties. The traditional statistical distribution such as Rayleigh with exponent tails cannot reflect the statistical characteristics of GF-3 SAR images [3, 4]. We use the heavy-tailed Rayleigh distribution to describe the characteristics of GF-3 SAR images [5]. The heavy-tailed Rayleigh distribution can accurately describe the sharp-parked and heavy-tailed statistical characteristics of the GF-3 SAR image when compared to the conventional Rayleigh, Weibull, and K distributions [6, 7]. Additionally, the maximum a posteriori (MAP) despeckling filter based on the heavy-tailed prior distribution is provided for GF-3 SAR images [8]. It can be seen that the MAP filter using heavy-tailed prior distribution leads to good performance, i.e., a better balance between speckle suppression and preservation of edges and point targets. This is attributed to the use of sharp-parked and heavy-tailed prior distribution. From the above experiments, we can conclude that the GF-3 SAR images are statistically sharp-parked and heavy-tailed, which is useful for the further processing of GF-3 SAR images.

2   GF-3 SAR Image

The GF-3 satellite has become the highest resolution satellite system in the world with the C-band multi-polarized SAR satellite. And the SAR satellite has the 12 imaging modes. The GF-3 satellite is widely applied to the ocean, disaster reduction, water conservancy, meteorology, and many other fields [9, 10]. It can provide users with long-term stable data support service. The study of the characteristics of GF-3 images is significant for practical application. GF-3 SAR image is shown in Fig. 1 and its parameters are listed in Table 1.

Fig. 1. GF-3 SAR image

Table 1. The parameters of GF-3 SAR image

No.	Imaging model	Polarization	Resolution	Imaging position
Figure 1	FSI	HH	5 m	E113.4 N34.7

3   Impulse Analysis of GF-3 SAR Image

Figure 2 shows the GF-3 SAR image. Figure 3 shows the sample of Fig. 2a. It can be seen that the GF-3 SAR images have the impulse. Obviously, GF-3 SAR image has sharp-peaked properties. The modeling results of Rayleigh distribution are shown in Fig. 4, including linear scale and logarithmic scale. Figure 4a shows the modeling result based on the traditional Rayleigh distribution [11, 12]. It can be seen that GF-3 SAR image has sharp-peaked properties and the traditional Rayleigh distribution cannot reflect the sharp peak of GF-3 SAR images. We can see that GF-3 SAR images have heavy-tailed statistical characteristics from Fig. 4b. Therefore, GF-3 SAR images have sharp-parked and heavy-tailed statistical characteristics. And GF-3 SAR images should be accurately described by an impulsive special distribution.

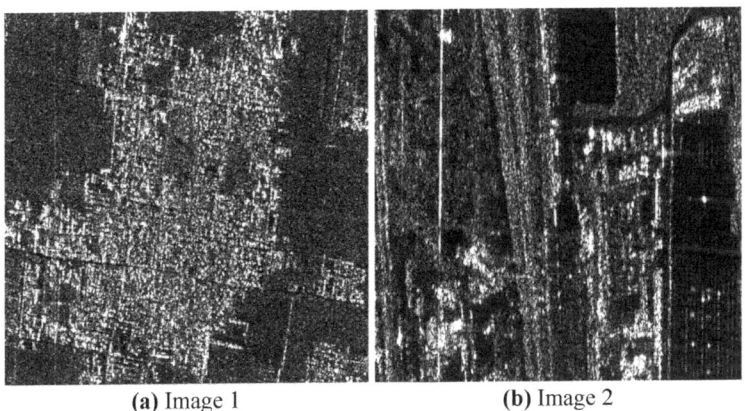

(a) Image 1   (b) Image 2

Fig. 2. GF-3 SAR image

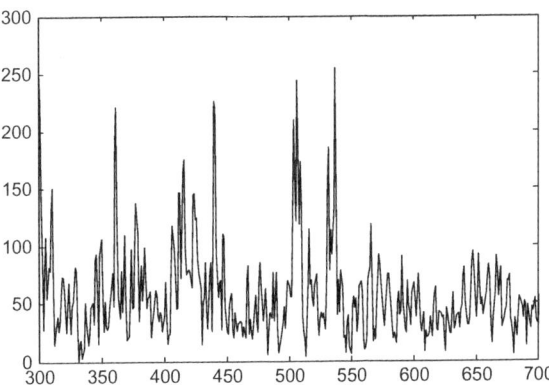

Fig. 3. Image sample corresponding to Fig. 2a

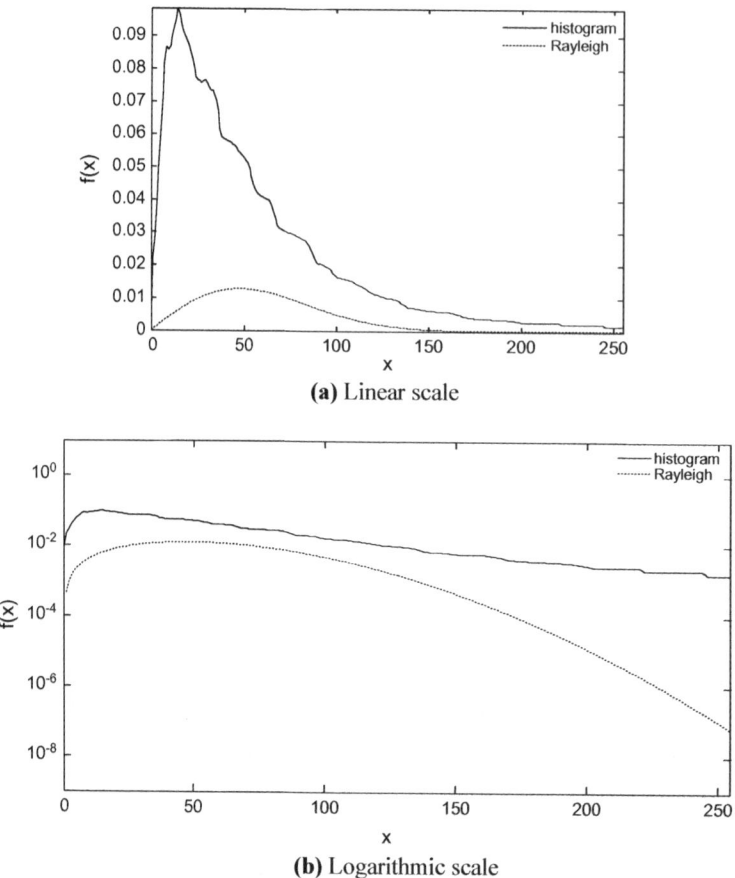

(a) Linear scale

(b) Logarithmic scale

Fig. 4. Modeling results of Rayleigh distribution of Fig. 2b

4 Modeling Experiment Based on Heavy-Tailed Rayleigh Distribution

According to the generalized central limit theorem, the stable distribution is the limiting distribution for sums of independent and identically distributed random variables in general case, and the Gaussian distribution is just a special case of the stable distribution. The stable distribution has algebraic tails, so it can reflect the sharp-peaked and heavy-tailed statistical properties of the impulsive signal and noise. Similar to stable distribution, heavy-tailed distributions also have algebraic tails. So heavy-tailed distributions also have high-peak and heavy-tail [7]. Consequently, heavy-tailed Rayleigh distribution is an accurate modeling tool for impulse signals and can accurately describe the statistical characteristics of GF-3 SAR images [13].

The probability density function (pdf) of the heavy-tailed Rayleigh distributions is given as follows [6, 7].

$$f_A(x) = x \int_0^\infty \rho e^{-\gamma \rho^\alpha} J_0(\rho x) d\rho \tag{1}$$

where α $(0 < \alpha < 2)$ is characteristic index, γ $(\gamma > 0)$ is scale parameter. $J(\cdot)$ is zero-order Bessel function of the first kind. The heavy-tailed Rayleigh distribution has the analytic pdf when the characteristic index $\alpha = 1$ and $\alpha = 2$.

Figure 5a shows the pdf of the heavy-tailed Rayleigh distributions using different values of α. Figure 5b shows the tail. Obviously. The heavy-tailed Rayleigh distribution has sharp-peaked and heavy-tailed statistical properties. And if the value of α is smaller, the sharp-peaked and heavy-tailed statistical property is clear.

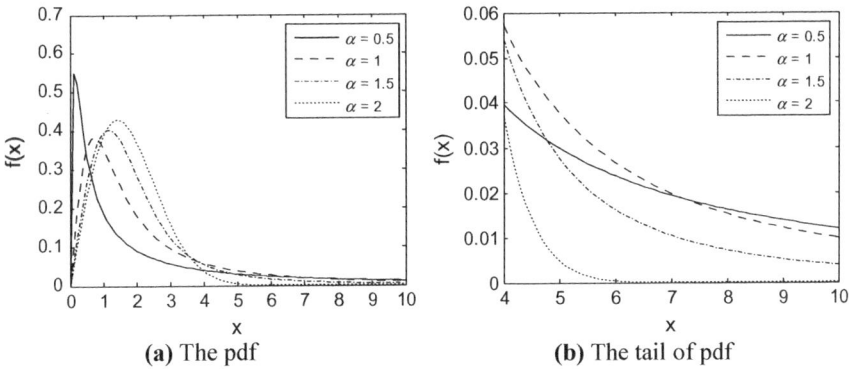

(a) The pdf   (b) The tail of pdf

Fig. 5. The pdf of the heavy-tailed Rayleigh distributions

Parameter estimation is a key issue to achieve accurate modeling of high-resolution GF-3 SAR images. The parameters of the heavy-tailed Rayleigh distribution were estimated by using the log-cumulant estimators [7, 14].

The log-cumulant estimator is given by [7]:

$$\tilde{k}_1 = C_e \left(\frac{1}{\alpha} - 1 \right) + \frac{\ln \gamma}{\alpha} + \ln 2 \tag{2}$$

$$\tilde{k}_2 = \frac{\pi^2}{6\alpha^2}, \tag{3}$$

where C_e is Euler's constant, γ is scale parameter and can be calculated from formula (2), and α is characteristic exponent and calculated from formula (3).

We model GF-3 SAR image to verify its characteristics. Figure 6 shows the modeling image. Figure 7 shows the GF-3 SAR image modeling results. Obviously, the Rayleigh distribution cannot be used to model GF-3 SAR amplitude images. Weibull distributions and K distributions have the heavy-tailed characteristics, but they cannot reflect the sharp-peaked characteristics. Heavy-tailed Rayleigh distribution has sharp-parked and heavy-tailed statistical characteristics. Therefore, heavy-tailed Rayleigh distribution is able to accurately describe characteristics of the GF-3 SAR image compared with the Rayleigh distribution, Weibull distributions, and K distributions [13, 15]. This means that GF-3 SAR image has algebraic tails.

Some quantitative measures and parameter estimation are provided in Table 2. The K-S probability is used to describe the proximity of a distribution to the statistical distribution of the image. If the value of K-S is bigger, this means that the distribution is closer to the statistical distribution of the image. As shown in Table 2, compared with other distributions, the heavy-tailed Rayleigh distribution can reflect the sharp-peaked and heavy-tailed characteristics of GF-3 SAR images. Therefore, we verify that the GF-3 SAR images own sharp-peaked and heavy-tailed statistical characteristics.

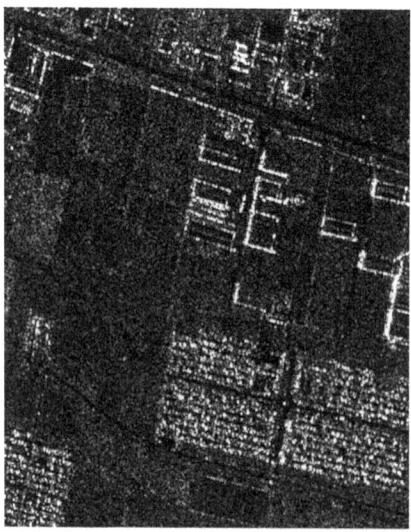

Fig. 6. Modeling image

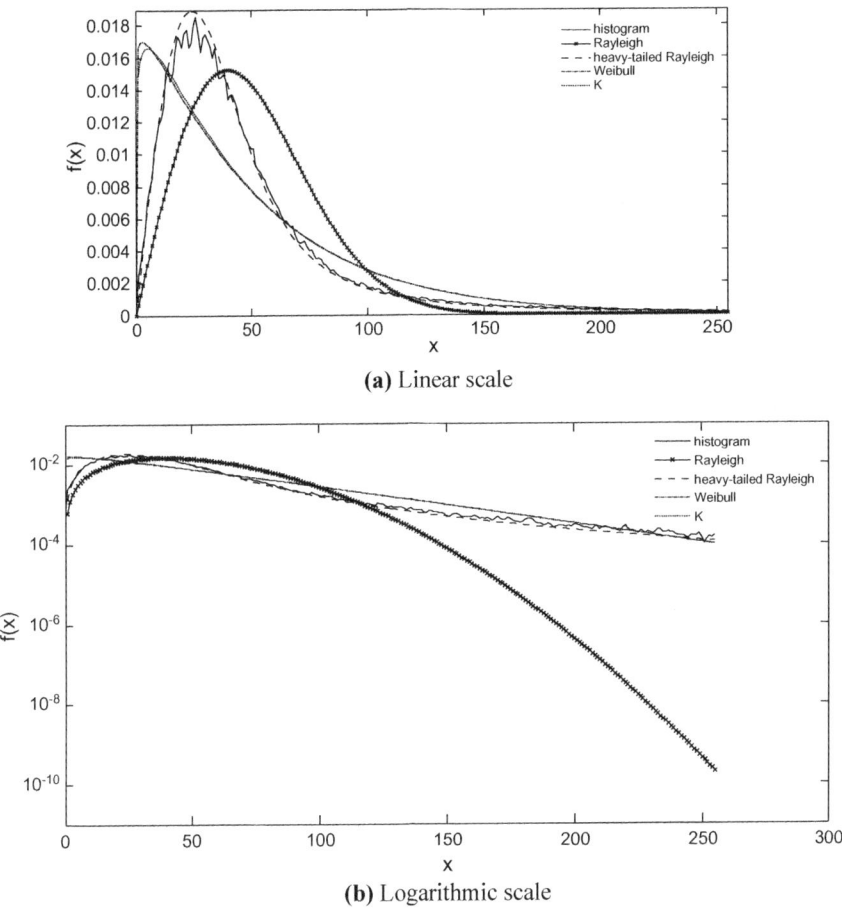

(a) Linear scale

(b) Logarithmic scale

Fig. 7. Modeling results corresponding to Fig. 6

Table 2. Quantitative measures and K-S probability for the modeling results of Fig. 7

	Parameter estimation results	K-S probability
Rayleigh distribution	$\hat{\gamma} = 791.9999$	0.8468
Heavy-tailed Rayleigh distribution	$\hat{\alpha} = 1.4917$	0.9755
	$\hat{\gamma} = 95.7864$	
Weibull distribution	$\hat{c} = 1.0506$	0.9009
	$\hat{b} = 50.8690$	
K distribution	$\hat{v} = 0.5690$	0.9038
	$\hat{b} = 0.0001199$	

5 MAP Despeckling Experiment Based on Heavy-Tailed Distribution

The general form of the MAP equation is given by [16, 17]

$$-\frac{1}{X} + \frac{\partial \ln f_F(F)}{\partial X} + \frac{\partial \ln f_X(X)}{\partial X}\bigg|_{X=\hat{X}_{MAP}} = 0, \tag{4}$$

where X is the real image, F is speckle. f_F and f_X are probability density functions of F and X respectively, $\hat{X}_{MAP}$ is MAP estimation of real image. It is obvious that the distributions for the speckle and the true image are required for constructing a MAP equation. Recalling the derivation of heavy-tailed Rayleigh distribution, the MAP despeckling experiment based on heavy-tailed distribution equation is constructed according to (4).

Heavy-tailed Rayleigh distribution do not have pdf except for the characteristic indexes $\alpha = 1$ and $\alpha = 2$. In this case, MAP despeckling experiment based on heavy-tailed Rayleigh distribution also do not have analytical formula except for the characteristic indexes $\alpha = 1$ and $\alpha = 2$. Heavy-tailed Rayleigh distribution becomes traditional Rayleigh distribution when $\alpha = 2$. So we select the cases with $\alpha = 1$ as an example.

where $\alpha = 1$, the MAP filtering equation is given by [16]

$$(2 + 2L)\hat{X}^2 + (2L\gamma^2 - 2LY)\hat{X} - 2L\gamma^2 Y = 0, \tag{5}$$

where L is the image looks, $\hat{X}$ is MAP estimation of real image, and Y denotes the observed image.

The simplified parameter estimation of the heavy-tailed distribution of the real image is given by [16, 17]

$$\hat{\gamma} = e^{\left[\hat{\tilde{k}}_{Y(1)} - 2\ln 2 - \psi(L) + \ln L\right]} \tag{6}$$

Here, $\hat{\tilde{k}}_{Y(1)}$ is the estimated values of the first order logarithmic cumulant of Y, $\tilde{k}_{Y(1)}$ can be estimated empirically from the observed image according to [7], and $\psi(\cdot)$ is Digamma function.

Figure 8 shows the despeckling experiment results. Figure 8a shows GF-3 SAR image. Figure 8b shows MAP filter based on heavy-tailed distribution. The despeckling results of the classical Lee filter [18, 19] and Gamma MAP filter [20] are provided for comparison. It is obvious that the MAP filter based on heavy-tailed prior distribution preserves edge information and point targets well and can suppress the speckle effectively in homogeneous region. However, the Lee filter and Gamma MAP filter cannot preserve the edge information and point well. Despeckling results of MAP filter based on heavy-tailed distribution are better than other methods [21]. This good performance is achieved from the use of the heavy-tailed distributions that describe the sharp-peaked and heavy-tailed characteristics of GF-3 SAR images. This means that the GF-3 SAR images are actually statistically sharp-peaked and heavy-tailed.

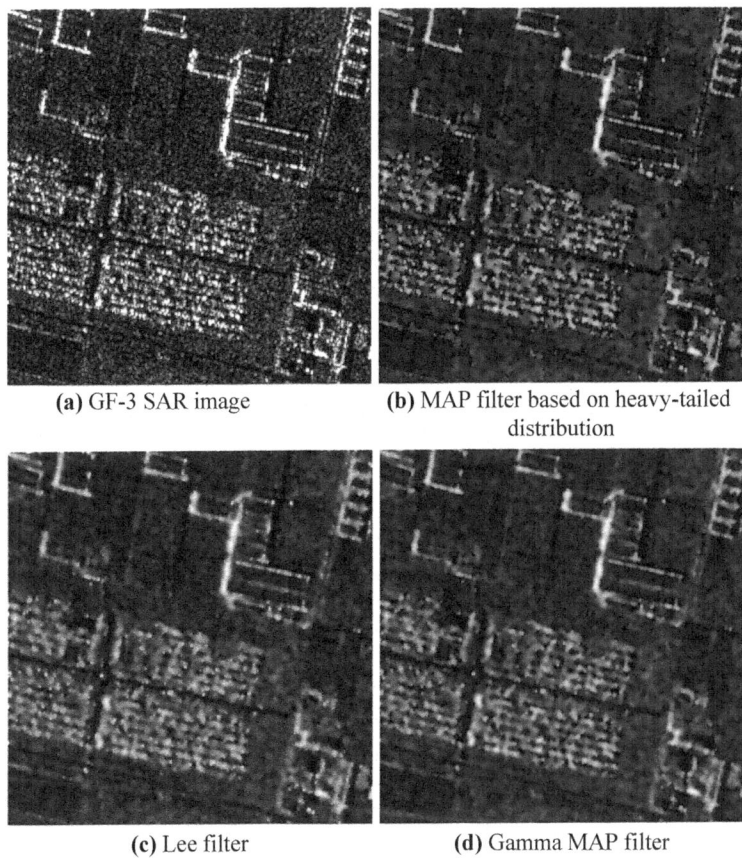

(a) GF-3 SAR image

(b) MAP filter based on heavy-tailed distribution

(c) Lee filter

(d) Gamma MAP filter

Fig. 8. Despeckling results of GF-3 SAR image

6 Conclusions

GF-3 satellite provides users with the long-time stable data support services. So the study of GF-3 image characteristics has become an important issue. We analyze characteristics of GF-3 SAR images from the image sample and statistical distribution of the image. It can be seen that GF-3 SAR images have sharp-peaked and heavy-tailed statistical properties. Furthermore, we verify the special characteristics of GF-3 SAR images based on related experiments including image modeling and MAP despeckling. The modeling experiments demonstrate that the heavy-tailed Rayleigh distribution is a better model for the GF-3 SAR images when compared to the conventional Rayleigh, Weibull, and K distributions. This means that the GF-3 SAR images own the obvious sharp-peaked and heavy-tailed statistical characteristics. Additionally, the MAP filter using the heavy-tailed prior distribution leads to a better balance between speckle suppression and preservation of edges and points in comparison with the Lee filter and Gamma MAP filter. This can be attributed to the choice of heavy-tailed distributions with sharp peak and heavy tail, which accurately corresponds to the real characteristics of GF-3 SAR images. In a word, from the above experiments, we can conclude that the GF-3 SAR images actually own the sharp-peaked and heavy-tailed characteristics. In the further research, we can apply these special statistical characteristics to produce new processing methods such as despeckling and classification that is appropriate for the GF-3 SAR images.

References

1. Jiayin Liu, Xiaolan Qiu, Bing Han, Dengjun Xiao. Study on Geo-Location of Sliding Spotlight Mode of GF-3 Satellite [C]. Proceedings of 5th IEEE Asia-Pacific Conference on Synthetic Aperture Radar, Singapore, 2015: 417–420.
2. Yujuan Guo, Erxue Chen, Ying Guo, Zengyuan Li, Chonggui Li, Kunpeng Xu. Deep Highway Unit Network for Land Cover Type Classification with GF-3 SAR Imagery [C]. 2017, 1–6.
3. Oliver C, Quegan S. Understanding Synthetic Aperture Radar Images [M]. Boston, MA: Artech House, 1998 New York: Academic, 1963, 271–350.
4. Z. Sun, Y. Song. Structural information detection based filter for GF-3 SAR images ISPRS - International Archives of the Photogrammetry, Remote Sensing and Spatial Information Sciences, 2018.
5. Brahimi B, Meraghni D, Necir A. Gaussian Approximation to the Extreme Value Index Estimator of a Heavy-tailed Distribution Under Random Censoring [J]. Mathematical Methods of Statistics, 2015, 24(4): 266–279.
6. Achim A, Kuruoglu E E, Zerubia J. SAR Image Filtering Based on the Heavy-tailed Rayleigh Model [J]. IEEE Transactions on Image Processing A Publication of the IEEE Signal Processing Society, 2006, 2686–2693.
7. Sun Z G, Han C Z. Modeling High-resolution Synthetic Aperture Radar Images with Heavy-tailed Distributions [J]. Phys. Sin. 2010, 59(02): 998–1008.
8. Zhao Y, Liu J G, Zhang B, et al. Adaptive Total Variation Regularization Based SAR Image Despeckling and Despeckling Evaluation Index [J]. IEEE Transactions on Geoscience & Remote Sensing, 2015, 53(5): 2765–2774.
9. Xuefei Zhang, Hong Zhang, Chao Wang. Water-change detection with Chinese GAOFEN-3 simulated compact polarimetric SAR images [C]. Proceedings of 2017 SAR in Big Data Era: Models, Methods and Applications, Beijing, China: 1–4.
10. Jiayin Liu, Xiaolan Qiu, Wen Hong. Automated ortho-rectified SAR image of GF-3 Satellite Using Reverse-range-Doppler Method [C]. Proceedings of International Geoscience and Remote Sensing Symposium, Beijing, China, 2016: 4445–4448.
11. Forbes C, Evans M, Hastings N, et al. Rayleigh Distribution [M]// Statistical Distributions, Fourth Edition. John Wiley & Sons, Inc. 2016:1416–1416.
12. Dey S, Dey T, Kundu D. Two-Parameter Rayleigh Distribution: Different Methods of Estimation [J]. American Journal of Mathematical & Management Sciences, 2014, 33(1): 55–74.
13. Achim A M, Kuruoglu E E, Zerubia J. Maximum a Posterior Estimation of Radar Cross Section in SAR Images Using the Heavy-Tailed Rayleigh Model [C]// 2005, European. IEEE, 2015: 1–4.
14. Liu, Tao, HaoGui Cui, Tao Mao, ZeMin Xi, and Jun Gao. Modeling multilook polarimetric SAR images with heavy-tailed rayleigh distribution and novel estimation based on matrix log-cumulants, Science China Information Sciences, 2013.

15. Sportouche H, Nicolas J M, Tupin F. Mimic Capacity of Fisher and Generalized Gamma Distributions for High-Resolution SAR Image Statistical Modeling [J]. IEEE Journal of Selected Topics in Applied Earth Observations & Remote Sensing, 2017, 10(12): 5695–5711.
16. Sun Z G. Gamma-Distributed Maximum A Posteriori Despeckling Algorithm of High-Resolution Synthetic Aperture Radar Images [J]. Phys. Sin. 2013, 62(18): 81–86.
17. Mahapatra D K, Ray S S, Roy L P. Despeckling of SAR Clutter Amplitude Data Using G0-MAP filter [C]// International Conference on Signal Processing and Communications. IEEE, 2016: 1–5.
18. Yommy A S, Liu R, Onuh S O, et al. SAR image despeckling and compression using K-nearest neighbor based lee filter and wavelet [C]// International Congress on Image and Signal Processing. IEEE, 2016: 158–167.
19. Lee J.S. Digital Image Enhancement and Noise Filtering by Use of Local Statistics [J]. IEEE Trans. Pattern Anal. Machine Intell. 1980, Mar, 2(2): 165–168.
20. Beauchemin, M., K. P. B. Thomson, and G. Edwards. Optimization of the Gamma-Gamma MAP filter for SAR image clutters [J]. International Journal of Remote Sensing 1996, 1063–1067.
21. Xiang Y, Wang F, You H. An Automatic and Novel SAR Image Registration Algorithm: A Case Study of the Chinese GF-3 Satellite [J]. Sensors, 2018, 18(2): 672–694.